A MATHEMATICAL THEORY OF GLOBAL PROGRAM OPTIMIZATION

Prentice-Hall
Series in Automatic Computation

George Forsythe, editor

AHO, editor, *Currents in the Theory of Computing*
AHO AND ULLMAN, *Theory of Parsing, Translation, and Compiling,*
Volume I: *Parsing;* Volume II: *Compiling*
ANDREE, *Computer Programming: Techniques, Analysis, and Mathematics*
ANSELONE, *Collectively Compact Operator Approximation Theory and Applications to Integral Equations*
ARBIB, *Theories of Abstract Automata*
BATES AND DOUGLAS, *Programming Language/One*, 2nd ed.
BLUMENTHAL, *Management Information Systems*
BRENT, *Algorithms for Minimization without Derivatives*
BRINCH HANSEN, *Operating System Principles*
COFFMAN AND DENNING, *Operating-Systems Theory*
CRESS, et al., *FORTRAN IV with WATFOR and WATFIV*
DANIEL, *The Approximate Minimization of Functionals*
DEO, *Graph Theory with Applications to Engineering and Computer Science*
DESMONDE, *Computers and Their Uses*, 2nd ed.
DESMONDE, *Real-Time Data Processing Systems*
DRUMMOND, *Evaluation and Measurement Techniques for Digital Computer Systems*
EVANS, et al., *Simulation Using Digital Computers*
FIKE, *Computer Evaluation of Mathematical Functions*
FIKE, *PL/1 for Scientific Programers*
FORSYTHE AND MOLER, *Computer Solution of Linear Algebraic Systems*
GAUTHIER AND PONTO, *Designing Systems Programs*
GEAR, *Numerical Initial Value Problems in Ordinary Differential Equations*
GOLDEN, *FORTRAN IV Programming and Computing*
GOLDEN AND LEICHUS, *IBM/360 Programming and Computing*
GORDON, *System Simulation*
HARTMANIS AND STEARNS, *Algebraic Structure Theory of Sequential Machines*
HULL, *Introduction to Computing*
JACOBY, et al., *Iterative Methods for Nonlinear Optimization Problems*
JOHNSON, *System Structure in Data, Programs, and Computers*
KANTER, *The Computer and the Executive*
KIVIAT, et al., *The SIMSCRIPT II Programming Language*
LORIN, *Parallelism in Hardware and Software: Real and Apparent Concurrency*
LOUDEN AND LEDIN, *Programming the IBM 1130*, 2nd ed.
MARTIN, *Design of Man-Computer Dialogues*
MARTIN, *Design of Real-Time Computer Systems*
MARTIN, *Future Developments in Telecommunications*
MARTIN, *Programming Real-Time Computing Systems*
MARTIN, *Security, Accuracy, and Privacy in Computer Systems*
MARTIN, *Systems Analysis for Data Transmission*
MARTIN, *Telecommunications and the Computer*
MARTIN, *Teleprocessing Network Organization*

MARTIN AND NORMAN, *The Computerized Society*
MATHISON AND WALKER, *Computers and Telecommunications: Issues in Public Policy*
MCKEEMAN, et al., *A Compiler Generator*
MEYERS, *Time-Sharing Computation in the Social Sciences*
MINSKY, *Computation: Finite and Infinite Machines*
NIEVERGELT et al., *Computer Approaches to Mathematical Problems*
PLANE AND MCMILLAN, *Discrete Optimization: Integer Programming and Network Analysis for Management Decisions*
PRITSKER AND KIVIAT, *Simulation with GASP II: a FORTRAN-Based Simulation Language*
PYLYSHYN, editor, *Perspectives on the Computer Revolution*
RICH, *Internal Sorting Methods: Illustrated with PL/1 Program*
RUSTIN, editor, *Algorithm Specification*
RUSTIN, editor, *Computer Networks*
RUSTIN, editor, *Data Base Systems*
BUSTIN, editor, *Debugging Techniques in Large Systems*
RUSTIN, editor, *Design and Optimization of Compilers*
RUSTIN, editor, *Formal Semantics of Programming Languages*
SACKMAN AND CITRENBAUM, editors, *On-line Planning: Towards Creative Problem-Solving*
SALTON, editor, *The SMART Retrieval System: Experiments in Automatic Document Processing*
SAMMET, *Programming Languages: History and Fundamentals*
SCHAEFER, *A Mathematical Theory of Global Program Optimization*
SCHULTZ, *Spline Analysis*
SCHWARZ, et al., *Numerical Analysis of Symmetric Matrices*
SHERMAN, *Techniques in Computer Programming*
SIMON AND SIKLOSSY, *Representation and Meaning: Experiments with Information Processing Systems*
STERBENZ, *Floating-Point Computation*
STERLING AND POLLACK, *Introduction to Statistical Data Processing*
STOUTEMYER, *PL/1 Programming for Engineering and Science*
STRANG AND FIX, *An Analysis of the Finite Element Method*
STROUD, *Approximate Calculation of Multiple Integrals*
TAVISS, editor, *The Computer Impact*
TRAUB, *Iterative Methods for the Solution of Polynomial Equations*
UHR, *Pattern Recognition, Learning, and Thought*
VAN TASSEL, *Computer Security Management*
VARGA, *Matrix Iterative Analysis*
WAITE, *Implementing Software for Non-Numeric Application*
WILKINSON, *Rounding Errors in Algebraic Processes*
WIRTH, *Systematic Programming: An Introduction*

A MATHEMATICAL THEORY OF GLOBAL PROGRAM OPTIMIZATION

MARVIN SCHAEFER

System Development Corporation

PRENTICE-HALL, INC.

ENGLEWOOD CLIFFS, N.J.

Library of Congress Cataloging in Publication Data

SCHAEFER, MARVIN,
A mathematical theory of global program optimization.

(Prentice-Hall series in automatic computation)
Bibliography: p.
1. Compiling (Electronic computers) 2. Mathematical optimization. I. Title.
QA76.6.S38 001.6′425 72-13785
ISBN 0-13-561662-X

10 9 8 7 6 5 4 3 2 1

Printed in the United States of America.

PRENTICE-HALL INTERNATIONAL, INC., *London*
PRENTICE-HALL OF AUSTRALIA, PTY. LTD., *Sydney*
PRENTICE-HALL OF CANADA, LTD., *Toronto*
PRENTICE-HALL OF INDIA PRIVATE LIMITED, *New Delhi*
PRENTICE-HALL OF JAPAN, INC., *Tokyo*

To my wife, Mary Alice

INTRODUCTION

With the introduction of algebraic programming languages and their compilers, there arose a great concern over the loss of efficiency (speed of execution) of compiler-generated assembly code as compared to handwritten code. Early compiler writers attempted to respond to this concern by writing compilers that would produce efficient code. So great was this concern for efficiency that some early languages, e.g., FORTRAN I, included linguistic features designed to instruct the compiler in various optimizations, including branching decisions.

As algebraic languages became more popular, a relatively standard set of algorithms for optimizing arithmetic expressions became common to most compilers. With languages like JOVIAL, programmers could "control" the optimization of certain segments of their programs by permitting sections of "direct code," i.e., assembly language, along with the higher-level statements. Thus, some programmers who might have been forced into writing their entire program in assembly language because of timing considerations were enabled to write only part of the code in assembly language. This "solution" did not add to the readability of programs, nor to the general accessibility of JOVIAL to programmers who were unfamiliar with the machine for which they were coding.

However efficient the code appeared to be, at best only one or two expressions on either side of a given expression were considered in the optimization. Any effect previous code may have had on the value of the expression —or any effect the expression might have on subsequent code—was not considered except within this limited, or "local," range.

All was not dismal, however. Many programs are only run a few times and are sufficiently short that speed is not a significant consideration. Besides, it has been argued, computers are getting much faster and applications are sufficiently complex to necessitate the more concise notation of higher-level algebraic languages; languages have become the trend in programming, and whatever optimization compilers can provide is appreciated, but not essential.

This brief history ignores the many gains in optimization in compilers, and does not touch on the problem of conserving space, which is as serious a problem as speed but which has not received the attention it deserves. It does, however, reflect the continuing attitude that truly efficient code must be handwritten assembly language, and that the only programs for which this code need be written are those that are executed frequently—namely, systems programs, utility programs, and real-time programs.

The common characteristic of these early attempts at optimization was that they examined the program on a statement-by-statement basis, treating code only in its local context. What was needed for more powerful optimization was a systematic emulation of the optimization techniques of assembly language programmers: examining statements in their *global* relationships with statements throughout the entire program.

There exist several meaningful efforts at developing global optimization techniques. F. E. Allen and J. Cocke of IBM have made major contributions by formalizing the approach to optimization. Their approach consisted of a directed-graph model of the program control flow and a partitioning of that graph into strongly connected subgraphs, corresponding to the program's loops, some of which were nested. With the assumption that the most deeply nested code was also the most frequently executed code, the optimization process started with code in the most deeply nested strongly connected subgraphs, working outward until the entire program graph has been covered.

The global optimizations performed by the experimental compilers of Allen and Cocke consisted of the elimination of common subexpressions, subsumption of constants, computation of invariant expressions outside the loops, and the reduction in strength of various operations in loops (e.g., transforming multiplications into iterative additions).

Actual compilers were produced for FORTRAN by C. P. Earnest of Computer Sciences Corporation for the UNIVAC 1107, 1108 and by E. S. Lowry and C. W. Medlock of IBM for S/360 computers. The general idea of isolating certain strongly connected subgraphs of the program flow graph and working from innermost nestings outward prevailed, although the partitionings of the graph were different from those of Allen and Cocke in that certain multi-entry strongly connected subgraphs were not treated as program loops. Essentially the same optimizations were carried out in both of these implementations, although different methods were employed.

These compilers have produced impressive results. It has been claimed that programs compiled by the IBM FORTRAN-(H) compiler run up to three times as fast, in 25 percent less space, than the same programs compiled under the FORTRAN-(G) compiler, with the compilation taking about 40 percent longer than the nonoptimizing compiler.

Global optimization requires a great deal of expensively obtained topological information from the program control flow graph. The requirement for this information is one of the major constraints to optimization. For example, isolating all strongly connected subgraphs is both time- and space-consuming. Hence, different partitions of the graph have been used in the industrial FORTRAN compilers.

Other time-consuming operations that tend to be abbreviated in practical optimizing compilers are the location of redundant subexpressions and the derivation of information about the flow of data in the program. An expression *e* located in two sections of a program is redundant if and only if all of the quantities involved in the computation of *e* maintain the same value between both computations, the determination of this fact is non-trivial. J. T. Schwartz, in conjunction with Allen, Cocke, and Earnest, has recently explored alternative partitionings of the program graph, and various topological transformations at lower cost. They have produced experimental compilers that treat multi-entry loops, derive execution frequency information, provide for additional operator strength reductions, and more closely examine data- and control-flow dependencies.

Close attention has also been paid to the optimization of machine register allocation, including the "bit-stripping" techniques of FORTRAN-(H), subroutine linkage, and storage maps and their relation to paging on 360-like computers. Some space vs. time optimizations have been considered, whereby iterative code might be transformed from a short loop to straight-line code with fewer branch-and-test operations, or where short subroutines are coded in-line rather than being branched to by subroutine calls.

It is clearly necessary that the equivalence of source code and optimized code be guaranteed; apparent equivalence is not sufficient. It is well known that not all computer operations follow the associative, commutative, and distributive laws of arithmetic, particularly when real and integral quantities are involved. Not so obvious are such machine-dependent variables as interrupt logic and the conditions that cause them; a program that runs to completion, however slowly, is more useful than an "optimized" version that either rapidly completes with erroneous results or aborts!

The requirement for this guarantee causes an almost paranoid conservatism on the part of investigators in the area. No optimization is performed unless a guarantee can be found assuring identical results with no possibility of causing machine interruptions. Mathematical proof of equivalence is necessary to the achievement of this goal, but not sufficient because of some of the poor fits between numerical approximations on a binary machine and the abstract mathematical structures employed for models.

It appears that the development of extremely good optimizing compilers, i.e., compilers which can optimize either space or speed as well as can human assembly language programmers, might eventually appear. Indeed, some mechanical checks can be made by the computer that are beyond the capability of mortals. Also, since optimizing compilers imitate the code-generation techniques of assembly language programmers, a greater number of special cases can be searched for and identified than would normally occur to any one programmer.

Some of the conservatism built into optimizing compilers can be circumvented if higher-level programming languages can once again be expanded to include features that would allow the programmer to pass on critical information to the compiler. These features might include execution- or branching-frequency information, and they might also allow the programmer to guarantee that, in certain cases, machine interrupts are impossible (e.g., overflow is not possible for an expression or a certain divisor can never equal 0).

This volume presents a unified treatment of several portions of the global optimization process. No attempt at completeness has been made, because of the rapid changes in the state of the art. It is assumed, for example, that the reader is aware of the more common local optimization techniques employed in good compilers for arithmetic and logical expressions.

It is also assumed that the reader has achieved that noetic entity called "mathematical maturity." Some prior exposure to abstract algebra and linear algebra is required in many of the proofs. Standard set-theoretic notation is employed throughout. Previous exposure to computer programming is helpful but not necessary.

Acknowledgements

I wish to thank the Advanced Research Projects Agency of the Department of Defense for its support of the research that has gone into this book, and Clark Weissman of System Development Corporation for his continued encouragement. I wish also to acknowledge the contributions made by my colleague Erwin Book and by John Cocke, C. P. Earnest, E. S. Lowry, and C. W. Medlock, whose techniques are incorporated and adapted in this presentation. Special appreciation is extended to Fran Allen, who was consulted on numerous occasions and tirelessly explained undocumented techniques and problems and to Donna Holder for her invaluable assistance in graph-theoretical modeling. Finally, to Val Schorre, Jean Ichbiah, Jonathan Goldstine and Seymour Ginsburg, who read proof and questioned the veracity of most of the theory and the relevance of the notation; to Debbie Miller and Shirley Wesson, who suffered the frustrating job of typing from illegible manuscript; and to my editor at Prentice-Hall, Judy Burke, I would like to express my thanks.

CONTENTS

INDEX OF SYMBOLS

A MATHEMATICAL THEORY OF GLOBAL PROGRAM OPTIMIZATION

PART I THEORETICAL FOUNDATIONS

Pure mathematics consists entirely of such asseverations as that, if such and such a proposition is true of anything, *then such and such another proposition is true of that thing. It is essential not to discuss whether the first proposition is really true and not to mention what the anything is of which it is supposed to be true . . . if our hypothesis is about* anything *and not about some one or more particular things, then our deductions constitute mathematics. Thus mathematics may be defined as the subject in which we never know what we are talking about, nor whether what we are saying is true.*

BERTRAND RUSSELL

OVERVIEW

To correctly identify those portions of a computer program which are susceptible to mechanical optimization, it is necessary to devise a methodology that will succinctly expose the complex interrelations between the flow of control and the flow of data during that program's execution. It is of special interest to isolate those segments of a program wherein it is possible to reach any one statement or data definition from any other statement or data definition, for it is within such program regions that we must currently hope to achieve our greatest gains in program efficiency. Unfortunately, it is also in these regions that we must be most careful about possible side effects from the optimization and their potential for destroying a previously working program.

The technology derived in this corpus is heavily based on the theory of finite directed graphs. We shall present an expeditious view of that theory in the following seven chapters. The presentation is oriented toward applications that appear in Part II of this treatise. Much of the theory is stated in a nonstandard form, so that it can immediately be exploited as required. Note, for example, that we shall not define the distance from a node of a graph to itself as zero, but as the length of the shortest nontrivial simple path connecting the node with itself. We shall also employ, in Chapter 6, the permanent, π, of the incidence matrix of the graph. While it is interesting to know that π is the permanent, we have refrained from the temptation to digress into combinatorial analysis and have simply made use of the required properties of π.

Proof is given for all major results and is sketched for the minor propositions. The reader may proceed directly to Chapters 4, 5, 6, and 7 if he feels his background in graph theory and set theory to be sufficiently strong.

1 PRELIMINARY NOTATION

1.1. GRAPH THEORY

We shall let uppercase Latin letters denote sets, lowercase Latin letters denote elements of sets, and Greek letters denote functions on sets. We shall not restrict ourselves to single-valued functions unless we specify that constraint.

We shall use the standard function notation conventions of set theory and algebra.

Let X be a set and let $\Gamma: X \longrightarrow 2^X$; i.e., Γ maps X into the set of all subsets of X.

By $\# X$ we mean the cardinality of the set X.

By convention we write $X \sim Y = X \cap \sim Y$ where $\sim Y$ is the complement of Y in some universal set $\mathfrak{U}$. $\mathfrak{U}$ will be clear from context.

We define a *graph* $G = (X, \Gamma)$ to be the pair consisting of the set X and the function Γ. We shall represent our graphs as planar sets. If $x, y \in X$ are two points such that $y \in \Gamma x$, we shall denote the relationship by joining the two points with a continuous curve with an arrowhead pointing from x to y. Hence a point $x \in X$ will be called a *vertex* or *node* of G, while the pair (x, y) with $y \in \Gamma x$ is called an *arc* of G.

Example

In Fig. 1.1 we show a graph G with $X = \{a, b, c, d, e, f, g, h\}$. Γ is defined with $\Gamma a = \{b, e\}$, $\Gamma b = \{c, d, f\}$, $\Gamma h = \varnothing$, etc.

A *subgraph* H of the graph $G = (X, \Gamma)$ is a graph of the form (A, Γ_A), where $A \subset X$ and Γ_A is defined as $\Gamma_A x = \Gamma x \cap A$. In Fig. 1.1, let $A = \{b, c, d\}$, $\Gamma_A b = \{c, d\}$, $\Gamma_A c = \varnothing$, and $\Gamma_A d = \{d\}$.

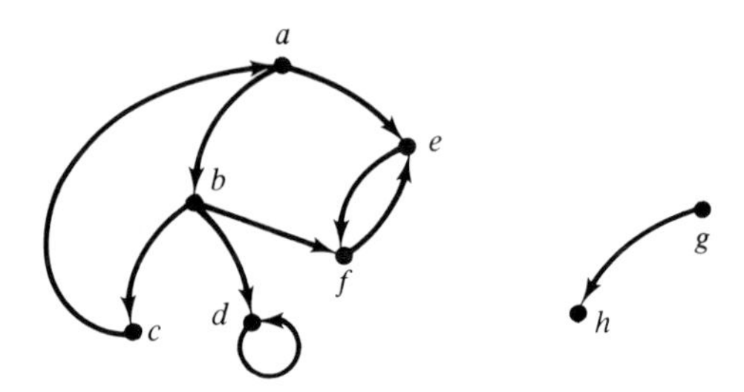

Figure 1.1

For an arc $u = (a, b)$, the vertex a is called the *initial vertex* of u, and b the *terminal vertex* of u. We also say that a is the *immediate predecessor* of b and that b is the *immediate successor* of a.

Two arcs u, v are *adjacent* if they are distinct and share a common vertex. In Fig. 1.1, (a, b) and (b, f) are adjacent, as are (b, d) and (d, d), etc.

The name *path* or *track* is given to a sequence $(u_1, u_2, \ldots, u_n)$ of arcs of a graph (X, Γ) such that for each node u_i in the sequence the terminal node of u_i is the initial node of u_{i+1}.

A path is *simple* if it does not use the same arc twice, and *composite* otherwise. A path is *elementary* if it does not meet the same vertex twice.

If a path μ meets, in turn, the vertices $x_1, x_2, \ldots, x_k, x_{k+1}$, we use the notation $\mu = [x_1, x_2, \ldots, x_k, x_{k+1}]$. A path using the same arcs as μ to get from x_2 to x_k is denoted as

$$\mu[x_2, x_k] = [x_2, \ldots, x_k]$$

A *circuit* or *cycle* is a path $\mu = [x_1, \ldots, x_k]$ in which $x_1 = x_k$. A circuit μ is called a *prime circuit* or *prime cycle* if there does not exist any circuit ν such that all of the vertices of ν are contained in μ; i.e., in this sense, μ contains no proper subcircuits.

The *length* $l(\mu)$ of a path $\mu = [x_1, \ldots, x_k]$ is the number of arcs in the sequence, and write $l(\mu) = k - 1$. A *loop* is a circuit of length 1, i.e., the single arc (x, x). In Fig. 1.1, $[d, d]$ is a loop and $[a, b, c]$ is a circuit.

We differentiate *arc* from *edge* as follows: An *arc* α exists between vertices x and y if $y \in \Gamma x$. An *edge* β exists between two vertices x and y if $y \in \Gamma x$ and $x \in \Gamma y$, i.e., if there exist arcs connecting x to y and y to x. Hence arcs have an inherent orientation and will be used in the sequel which deals with directed graph representations of program topology. It will suffice for our purposes to represent any number of arcs connecting x and y by a unique arc connecting x to y.

1.2. LIMITS OF SEQUENCES

It will be necessary at times to refer to the formal limit of a sequence of sets. In this discussion we shall always be using finite sets, so that any sequence of subsets of elements from the set X must necessarily contain a finite number of distinct terms, in fact no more than $2^{\#X}$.

We let $\{\alpha_\nu\}$, $\nu = 1, 2, \ldots$, be a sequence of subsets of X. We say that

$$\lim \alpha_\nu = \lim_{\nu \to \infty} \alpha_\nu$$

exists, and we write

$$\lim \alpha_\nu = \alpha$$

if, from some point N on, $\alpha_N = \alpha_{N+1} = \cdots$. Here, $\alpha = \alpha_N$.

2 WEAK ORDERING ASSOCIATED WITH A GRAPH

2.1. FUNCTIONAL NOTATION

We expand our functional notation as follows: Let $x \in X$. Then $\Gamma x \in 2^X$. We define $\Gamma^2, \Gamma^3, \ldots$ as

$$\Gamma^2 x = \Gamma(\Gamma x)$$
$$\Gamma^3 x = \Gamma(\Gamma^2 x) = \Gamma(\Gamma(\Gamma x))$$
$$\vdots$$

The *transitive closure* of Γ is a function $\hat{\Gamma}: X \longrightarrow 2^X$ defined by

$$\hat{\Gamma} x = \{x\} \cup \Gamma x \cup \Gamma^2 x \cup \cdots$$

The inverse of Γ is a function $\Gamma^{-1}: X \longrightarrow 2^X$ defined by

$$\Gamma^{-1} y = \{x \mid y \in \Gamma x\}$$

The transitive closure of Γ^{-1} is a function $\hat{\Gamma}^{-1}: X \rightarrow 2^X$ defined by

$$\hat{\Gamma}^{-1} x = \{x\} \cup \Gamma^{-1} x \cup \Gamma^{-1}(\Gamma^{-1} x) \cup \cdots.$$

If $B \subset X$, we write $\Gamma^{-1} B = \{x \mid \Gamma x \cap B \neq \varnothing\}$.

If $A \subset X$, we write $\Gamma^{\S} A = \{x \mid x \notin A$ and there exists $y \in A$ such that $x \in \Gamma y\} = \Gamma A \sim A$.

Hence

1. $\Gamma^3 x$ is the set of vertices which can be reached in *three* moves from vertex x.
2. $\Gamma^n x$ is the set of vertices which can be reached in n moves from vertex x.
3. $\hat{\Gamma} x$ is the set of vertices which can be reached from the vertex x (including x itself).
4. $\Gamma^{-1} A$ (where $A \subset X$) is the set of vertices which can, after one move, yield a vertex in the set A.
5. $\Gamma^{-n} A$ (where $A \subset X$) is the set of vertices which can, after n moves, yield a vertex in the set A.

2.2. CONNECTIVITY

A graph G is *connected* if for every $x, y \in G$ either $y \in \Gamma x$ or $x \in \Gamma y$. G is *strongly connected* if for every $x, y \in G$ both $y \in \hat{\Gamma} x$ and $x \in \hat{\Gamma} y$.

The graph of Fig. 2.1 is connected, but not strongly connected, since, e.g., while there is a path from a to g, there is no path from g to a. The graph in Fig. 2.2 is strongly connected as there is a path starting at any given node ending at any other node.

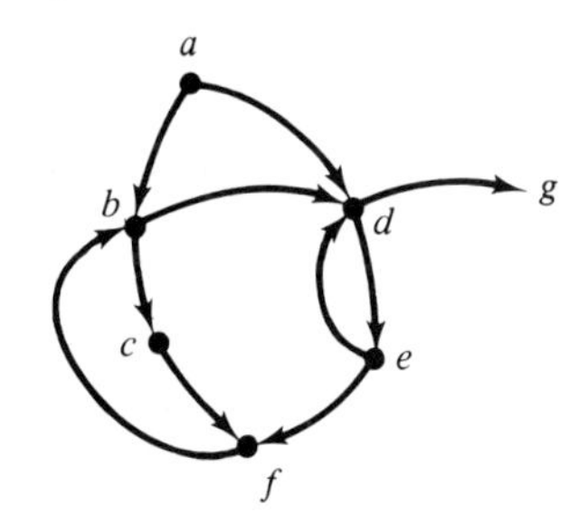

Figure 2.1

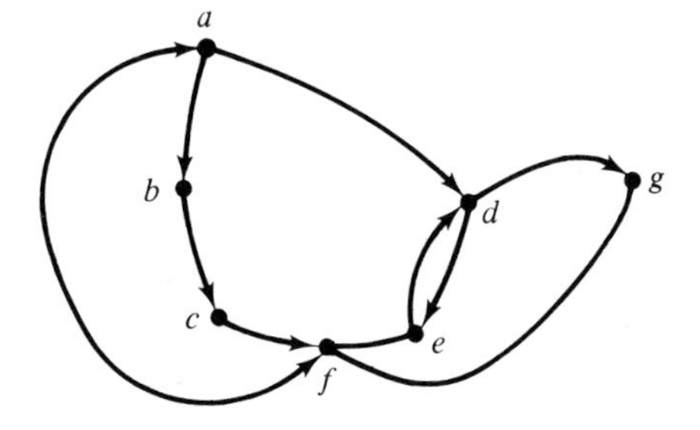

Figure 2.2

2.3. ORDER RELATIONS

2.3.1. Equivalence Relations

We shall now recall the notions of order relations. If $\equiv$ is a relation satisfying the properties

1. *Reflexivity:* $x \equiv x$ (for all $x \in X$),
2. *Symmetry:* $x \equiv y \Rightarrow y \equiv x$, and
3. *Transitivity:* $x \equiv y, y \equiv z \Rightarrow x \equiv z$,

then $\equiv$ is called an *equivalence relation.* The classes $\mathcal{C}x = \{y \mid y \equiv x\}$ form a partition of X into disjoint *equivalence classes.*

2.3.2. Partial Orderings

Let $\equiv$ be an equivalence relation on X. Then a relation $\leqslant$ satisfying

1. *Reflexivity:* $x \leqslant x$ (for all $x \in X$),
2. *Antisymmetry:* $x \leqslant y$ and $y \leqslant x \Rightarrow x \equiv y$ (where $\equiv$ is an equivalence relation on X), and
3. *Transitivity:* $x \leqslant y$ and $y \leqslant z \Rightarrow x \leqslant z$

will be called a *partial ordering on* X. If property 2′ is not satisfied, $\leqslant$ constitutes a *weak ordering.*

In the definition of partial order we assume that $\equiv$ is a preestablished equivalence relation on X. It is clear that for a given equivalence relation $\equiv$ on a set X there exist order relations which do not satisfy the above criteria. Alternatively, if we were to start with a given order relation and reword 2′ as

2″. $a \leqslant b$ and $b \leqslant a \Leftrightarrow a \sim b$,

we may show that $\sim$ is *the* equivalence relation induced by the partial order $\leqslant$,

(for since $a \leqslant a$, certainly also $a \leqslant a$ so that $a \sim a$. Also, from $a \sim b$ we obtain $a \leqslant b$ and $b \leqslant a$, i.e., $b \leqslant a$ and $a \leqslant b$, so that $b \sim a$. Finally, $a \sim b$ and $b \sim c$ give $a \leqslant b$ and $b \leqslant c$, implying that $a \sim c$).

2.3.3. Total Orderings

The law of trichotomy for an arbitrary relation $\leqslant$ requires that for all $x, y \in X$ either

1. $x \leqslant y$,
2. $x \equiv y$, or
3. $y \leqslant x$

is true. If a partial order $\leqslant$ satisfies the law of trichotomy, it is called a *total ordering* (or a *linear ordering*) on X.

A totally ordered subset C of a partially ordered set X is called a *chain.*

Example

We shall give an example of an order relation $\leqslant$ for an arbitrary graph $G = (X, \Gamma)$ by defining

$$\text{for } x, y \in X, x \leqslant y \text{ if } y \in \hat{\Gamma}x$$

and call x a *predecessor* of y. It is clear that properties 1 and 3 are satisfied, so that $\leqslant$ is a weak order of any graph.

Now if we have both $x \leqslant y$ and $y \leqslant x$, it follows that both $y \in \hat{\Gamma}x$ and $x \in \hat{\Gamma}y$. We conclude that either x and y coincide or that x and y lie on the same circuit. $x \equiv y$ is easily shown to satisfy the reflexivity, symmetry, and transitivity requirements of an equivalence relation.

The equivalence classes determined by $\equiv$ have characteristic properties for two extreme forms of directed graphs; viz., if $G = (X, \Gamma)$ is circuit-free, then there will be $\# X$ equivalence classes, while if G is strongly connected, there will be but one equivalence class. All other graphs are somewhere between these extremes.

When we have $x \equiv y$ in a circuit-free graph, we identify x with y and write $x = y$.

2.4. MAXIMAL AND MINIMAL ELEMENTS

If $x \leqslant y$ but not $x \equiv y$, we write $x \prec y$ and state the following: The vertex x is an *ancestor* of y. We may also write $y \succ x$, stating the following: The vertex y is a *descendant* of x.

If $B \subset X$, a vertex z which follows (precedes) all the vertices of B is called a *majorant* (*minorant*) of B; we write $z \succcurlyeq b$ $(b \in B)$ $[b \succcurlyeq z$ $(b \in B)]$.

If B contains an element which is a majorant (minorant) of B, this element is called a *maximum* (*minimum*) of B. If b and b' are two maxima of B, we have $b \leqslant b'$ and $b' \leqslant b$ so that $b \equiv b'$. Hence all maxima are equivalent to one another.

A set $B \subset X$ is a *basis* of the graph $G = (X, \Gamma)$ if

1. $b_1, b_2 \in B, b_1 \neq b_2 \Rightarrow$ neither $b_1 \leqslant b_2$, nor $b_2 \leqslant b_1$.
2. $x \notin B \Rightarrow$ there exists $b \in B$ such that $b \leqslant x$.

A graph is called *inductive* if every path $\mu = [x_1, x_2, \ldots]$ possesses a majorant, i.e., if a vertex z exists such that

$$z \geqslant x_n \qquad (n = 1, 2, \ldots)$$

All finite graphs are inductive.

We recall the following well-known result of set theory:

ZORN'S LEMMA

If T is partially ordered and each linearly ordered subset has an upper bound in T, then T contains at least one maximal element.

In the previous context we note that inductive graphs clearly possess linearly ordered paths which possess upper bounds (majorants) and may therefore rephrase the result as

ZORN'S LEMMA FOR INDUCTIVE GRAPHS

For every vertex x of an inductive graph (X, Γ) there exists a vertex z without descendants such that $z \geqslant x$ and a vertex q without ancestors such that $x \geqslant q$.

(A rigorous discussion of this result may be found in Chapter III of C. Berge, *Espaces topologiques et Fonctions multivoques*, No. III, Collection Universitaire de Mathématiques, Dunod, Paris, or in R. Croisot, M. L. Dubreil-Jacotin, and L. LeSieur, *Leçons sur la Théorie des Treillis*, Gauthier-Villars, Paris, 1953.)

The utility of the above results is generally exploited in practical problems concerning communication networks and the theory of games, as well as purely theoretical applications to the topological analysis of infinite graphs. We are endeavoring to integrate communication network theory into program optimization and include these results as a theoretical foundation upon which we hope eventually to be able to rely.

Much of the practical optimization we wish to perform will be applied to the strongly connected subgraphs of programs. The equivalence classes of these regions under $\leqslant$ collapse to include the entire region. In the next chapter we shall derive a more useful partial order for program graphs.

EXERCISES

1. Consider the graph representing the family tree of your family starting from your paternal grandfather.

a. Is it connected? If not, what modifications to the connectivity would be required to connect it?

b. Impose the following relation on the graph: If $y \in \hat{\Gamma}x$, then $x \sim y$, and we say that x is an ancestor of y. Which of the order relations are satisfied by the nodes of this graph? Is $\sim$ a total ordering? Partial ordering? Are there chains in the graph? If so, who is the maximal element?

2. Criticize the following "proof" that any relation $\sim$ which is both symmetric and transitive is automatically reflexive:

> From symmetry, we have $x \sim y \Rightarrow y \sim x$. Applying transitivity to $x \sim y$ and $y \sim x$, we obtain both $x \sim x$ and $y \sim y$.

3 DOMINANCE, PARTITIONS, AND INTERVALS OF GRAPHS

In Part II we shall be relating the theory of graphs to the global optimization of computer programs. By *optimization* we mean the reduction of some cost function (e.g., number of instructions, time required for execution, required computer store) to a minimum. We shall distinguish between global and local optimization in the usual sense; i.e., global optimization covers larger segments of a program than does local optimization and hence includes it. It is our feeling that sufficient literature is available on the techniques involved in local optimization that we may omit discussion of such techniques from this treatment.

3.1. THE CONTROL FLOW GRAPH

Basically, a computer program is a sequence of instructions which are usually executed consecutively. These instructions either compute quantities, make decisions, or transfer control to an instruction which is out of the normal execution sequence and hence to the instructions following it.

We shall represent the flow of control in computer programs with a finite directed graph.

DEFINITION **3.1**

A block is a maximal sequence of instructions having the property that whenever any one of the instructions in the sequence is executed, every instruction in the block is executed.

Since a program consists of a finite number of instructions, a block is also finite and is linearly ordered by the normal execution sequence of the computer. Hence there is a unique first instruction f in a block b and a unique final instruction l, which necessarily transfers control to some instruction i out of the normal execution sequence. This instruction, i, has the property that there is some other instruction l' which also transfers control to it, for if not, execution of i would entail prior execution of l so that i would be in the same block as l, contradicting the maximality of the block. Furthermore, i must be the first instruction in some block b', for if not, it would be possible to execute i without executing all the instructions in b'. Thus we obtain

PROPOSITION **3.2**

Let b be a block. Then there is a unique first instruction i in b. If i is not the first instruction in the program, then there exists a transfer instruction l such that l transfers control directly to i. Finally, the last instruction in each block b either transfers control to the first instruction of some block, possibly b itself, or is a terminal instruction in the program.

DEFINITION **3.3**

A *basic block* b' is a contiguous subsequence of instructions contained in a block b with the property that the first instruction of b' is the first instruction of b or that there exists an instruction in b which transfers to b', and the last instruction of b' is either the last instruction of b or a transfer to some instruction of b.

In terms of programming languages, a basic block typically has a first instruction possessing a label (explicit or implicit) and whose last instruction is the first transfer instruction following the label.

We let X be the set of (basic) blocks in a program P, and for two (basic) blocks $x, y \in X$ we say that $y \in \Gamma x$ if the last instruction in x is a transfer to (the first instruction of) y. With Γ thus defined, $G = (X, \Gamma)$ is called the *control flow graph of P*.

The choice of basic blocks over blocks is one of convenience; most programming languages include explicit language forms, transfer statements, labels, conditionals, etc., so that identification of basic blocks is far easier than the identification of maximal sequences of instructions constituting blocks. The main reason for using such partitioning into (basic) blocks is the compactification of a program graph into something less involved than a graph in which each instruction constitutes a node.

It is known that programs have entry points, and since G is a finite (and hence inductive) graph, Zorn's lemma guarantees that we may identify these ancestor-free vertices.†

We define *exit block* or *exit nodes* to be descendant-free nodes.‡

3.2. BACK DOMINANCE

We also define an ordering $\leq$, called *back dominance*. Let E be the set of entry nodes of G and let $e \in E$. Then we say that x *back-dominates* y with respect to e if for every simple path $\mu = [e, x_0, x_2, \ldots, y]$ we have $x \in \mu$. Clearly, x is an ancestor or predecessor of y if $x \leq y$. *Forward dominance* may be defined similarly in terms of exit vertices. All results that hold for back dominance hold (with obvious modification) for forward dominance.

For simplicity we shall assume that G has only one entry point e. (If not, we simply generate a node e which is ancestor-free and have it branch to each of the previous entry nodes.) Then every node is back-dominated by at least one node.

Example 3.4

In Fig. 3.1, we let e be the entry node of the graph G and let s and t be the exit nodes. e back-dominates all of G. s is back-dominated by e, l, and m; k by e and f; and t by e, g, and j.

PROPOSITION **3.5**

$\leq$ is a partial order.

Proof. For any $x \in X$ and any $\mu = [e, x_1, \ldots, x_n = x]$, we have $x \in \mu$. Hence $x \leq x$.

Next, if $x \leq y$ and $y \leq x$, we have for every $\mu = [e, x_1, x_2, \ldots, y]$ and every $\nu = \{e, y_1, y_2, \ldots, x]$ that $y \in \nu$ and $x \in \mu$. Therefore consider the set $M = \{\mu \mid \mu$ is a simple path from e to $y\}$.

Then, given $\mu \in M$, $\mu = [e, x_1, x_2, \ldots, x_i, x, x_{i+2}, \ldots, y]$, it must be that the path $\nu = \mu[e, x] \in M$. But then $y \in \nu$, i.e., $\nu = [e, x_1, x_2, \ldots, x_j, y, x_{j+2}, \ldots, x]$, which implies that μ is not a simple path, a contradiction. Hence $x = y$.

Finally, we show that if $x \leq y$ and $y \leq z$, then $x \leq z$. Let

$$M = \{\mu \mid \mu \text{ is a simple path from } e \text{ to } y\}$$
$$N = \{\nu \mid \nu \text{ is a simple path from } e \text{ to } z\}$$

†In some cases, a program cycles indefinitely through what is commonly referred to as its entrance. In those cases, the entry point referred to above is the initial linkage code generated by the compiler from programmer-generated information.

‡Similarly, programs may not have exit nodes.

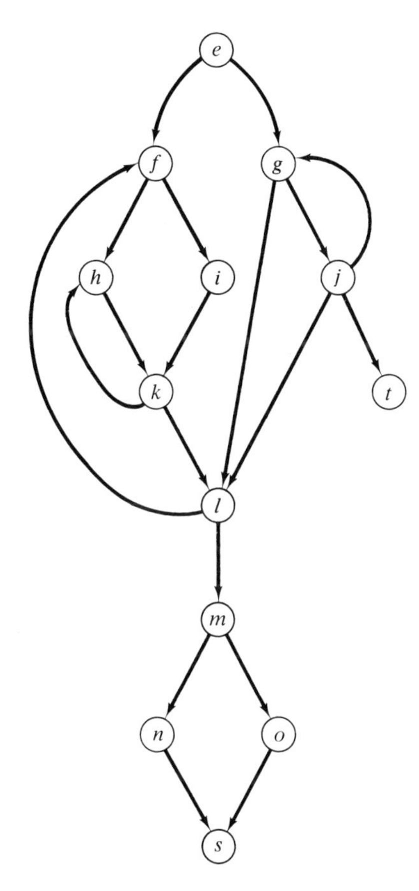

Figure 3.1

Then for any $\nu \in N$, $y \in \nu$ since $y \leq z$. Hence there is a $\mu \in M$ such that $\mu = \nu[e, y]$. Now since $x \leq y$, $x \in \mu$. But since $\mu \subset \nu$, $x \in \nu$, i.e., $x \leq z$. ▮

Since back dominance forms a partial order and we are dealing with finite graphs, each $x \in X$ has a finite chain $\mathfrak{B}_x$ of back dominators.† By Zorn's lemma, each chain must have a minimal element, viz., a vertex $y \in X$ which (properly) back-dominates x and is back-dominated by all other back dominators of x. Such a y is called the *immediate back dominator of* x.

THEOREM **3.6**

The set $\mathfrak{B}_x$ of back dominators of a vertex $x \in X$ is given by

$$\mathfrak{B}_x = \bigcap_{y \in \Gamma_x^{-1}} (\mathfrak{B}_y \cup \{y\})$$

Trivially, if $\Gamma^{-1}x = \{y\}$, then $\mathfrak{B}_x = \mathfrak{B}_y \cup \{y\}$.

†For practical reasons, we shall exclude x from $\mathfrak{B}_x$.

Proof. If $\Gamma^{-1}x = \{y\}$, then $y \leq x$ since y must lie on every path to x. If x has several predecessors y_i, then any back dominator x_y of y_i is a predecessor of x and hence must lie on some path to x. Hence those back dominators of the immediate predecessors of x which lie on every path from e to x back-dominate x. Finally, if x has any other back dominators, they must be included in every simple path from e to x and either already be included in each $\mathfrak{B}_{y_i}$ or be one of the y_i. This last possibility would occur if, for example, y and z are immediate predecessors of x while $z \in \hat{\Gamma}x$ (i.e., there is a circuit from x to itself including z); hence we also have $\mathfrak{B}_y \subset \mathfrak{B}_z$ and $y \in \mathfrak{B}_z$. ▮

Theorem 3.6 provides us with an iterative algorithm for finding the back dominators of each vertex of the program graph (X, Γ).

3.3. THE PARTITION INTO INTERVALS

DEFINITION **3.7**

We define an *interval* I with initial node (or head) h as the maximal subgraph (I, Γ_I) of (X, Γ) with the properties

1. $h \in I$.
2. If $x \in I$, $x \in \hat{\Gamma}h$.
3. $I \sim \{h\}$ is cycle-free.
4. If $x \in I \sim \{h\}$, $\Gamma^{-1}x \subset I$.

We alternately write $I(h)$ to mean an interval I with initial node h. Next we shall give an algorithm for constructing intervals.

Example

The intervals of Fig. 3.1 are $I(e) = \{e\}$, $I(f) = \{f, i\}$, $I(g) = \{g, j, t\}$, $I(h) = \{h\}$, $I(k) = \{k\}$, and $I(l) = \{l, m, n, o, s\}$.

ALGORITHM **3.8**

To construct a class $\mathcal{I}$ of intervals of (X, Γ), where $X = \{x_1 = e, x_2, \ldots, x_n\}$,

1. Let $y_1 = x_1$.
2. $I_{j1} = \{y_j\}$.
3. $I_{j,k+1} = I_{jk} \cup \{x \in \Gamma^{\S} I_{jk} \mid \Gamma^{-1}x \subseteq I_{jk}\}$.
4. $I_j = \lim_k I_{jk}$.

5. If $\Gamma^{\S}(I_1 \cup \cdots \cup I_j) \neq \varnothing$, let $y_{j+1} = x_l$, where

$$l = \min \{m \mid x_m \in \Gamma^{\S}(I_1 \cup \cdots \cup I_j)\}$$

6. Repeat steps 2, 3, 4, and 5.

Then $\mathcal{I} = I_1 \cup \cdots \cup I_j$.

The reader is invited to apply Algorithm 3.8 to the graph of Fig. 3.1 to ensure his comprehension of the interval concept.

THEOREM **3.9**

Algorithm 3.8 uniquely partitions $(\hat{\Gamma}e, \Gamma_{\hat{\Gamma}e})$ into disjoint intervals.

LEMMA **3.9.1**

An interval is produced by steps 2-4.

Proof. For any arbitrary $y_j = x_p \in X$, we show that I_j is an interval. Clearly, $y_j \in I_j$, so we show that properties 2, 3, and 4 of Definition 3.7 are satisfied by I_j.

If $x \in I_{jk}$ for some $k > 1$ while $x \notin I_{j,k-1}$, then $x \in \Gamma^{\S} I_{j,k-1}$ and $\Gamma^{-1}x \subseteq I_{j,k-1} \subseteq I_j$. Hence property 4 is satisfied, while by induction, property 2 is also satisfied.

Assume that $I_j \sim \{y_j\}$ is not cycle-free. Then there exist vertices $u_1, \ldots, u_m \in I_j \sim \{y_j\}$ such that $\mu = [u_1, \ldots, u_m]$ is a cycle of $I_j \sim \{y_j\}$, where $u_1 \in \Gamma u_m$. However, then none of the members of μ could be included in I_j, since if, e.g., u_1 were just considered at some stage k in step 3, we would have $u_m \in \Gamma^{-1} I_{j,k-1}$ so that $u_{m-1} \in \Gamma^{-1} I_{j,k-2}$, etc., so that $u_1 \in \Gamma^{-1} I_{j,k-m-1}$, a contradiction to $u_1 \notin I_{j,k-1}$.

Finally, the maximality of I_j is equivalent to showing the existence of the limit in step 4 which follows from the finiteness of X. ∎

LEMMA **3.9.2**

A given interval head y_j produces a unique interval I_j.

Proof. Assume the existence of two intervals I_j and I'_j with some vertex $x \in I_j \sim I'_j$ and $\Gamma^{-1}x \subset I_j \cap I'_j$. Then there must exist an integer n such that for all $k > n$, $x \in \Gamma^{\S} I'_{jk}$ while $\Gamma^{-1}x \not\subset I'_{jk}$, for otherwise x could be admitted to $I_{j,k+1}$ and hence eventually to I'_j.

Now since $\Gamma^{-1}x \subset I_j \cap I'_j$, we need conclude that there is an integer n' such that for all $k' > n'$

$$\Gamma^{-1}x \subset I_{jk'} \cap I'_{jk'}$$

Choose $k'' = \max(n, n')$. Then $\Gamma^{-1}x \subset I'_{jk''}$, contradicting the maximality of I'_j. ▮

LEMMA **3.9.3**

For any $j \geq 1$, any $x_i \in \Gamma^{\$}(I_1 \cup \cdots \cup I_j)$ becomes an interval head.

Proof. If $x_i \in \Gamma^{\$}(I_1 \cup \cdots \cup I_j)$, then $\#\Gamma^{-1}x_i \geq 2$, for if not, we would have $x_i \in \Gamma^{\$}I_m$ $(1 \leq m \leq j)$ and $\Gamma^{-1}x_i \subset I_m$ so that $\Gamma^{-1}x_i \subset I_{mk}$ for all k sufficiently large, implying $x_i \in I_m = \lim_k I_{mk}$, a contradiction. ▮

Proof of Theorem 3.9. By Lemma 3.9.1, I_1 is unique, and each $x_i \in \Gamma^{\$}I_1$ produces a unique interval I_i by Lemma 3.9.3, which is disjoint from all other intervals $I_{i'}$. Since (X, Γ) is finite, eventually

$$\Gamma^{\$}(I_1 \cup \cdots \cup I_j) = \varnothing$$

so that the algorithm terminates.

Since each $x_m \in I_1 \cup \cdots \cup I_j$ belongs to a unique I_i, it follows from the above that $x_m \in \hat{\Gamma}e$, i.e., that $\hat{\Gamma}e = I_1 \cup \cdots \cup I_j$. ▮

There may exist $x \in X \sim \hat{\Gamma}e$. Such x are isolated from the single-entry vertex e of (X, Γ) and probably represent a programmer error.

PROPOSITION **3.10**

Any interval I of G is contained within some interval $\hat{I} \in \mathscr{I}$, where $\mathscr{I}$ is the collection of intervals produced by Algorithm 3.8.

Proof. Let $I = I(h)$ so that h is the head of I. By Lemma 3.9.2, $I(h)$ is unique, and for any $x \in I(h) \sim \{h\}$, $\Gamma^{-1}x \subseteq I(h)$. Hence, if h heads any interval $I_h \in \mathscr{I}$, we have $I(h) = I_h$ and are done.

Therefore let $h \in I_j \in \mathscr{I}$, with $\Gamma^{-1}h \subseteq I_j$. This is possible since $X = \cup_{I_j \in \mathscr{I}} I_j$. Every $x \in I(h)$ is an element of I_j, since $x \in \hat{\Gamma}h \subseteq \hat{\Gamma}y_j$ and $\Gamma^{-1}x \subseteq I(h)$ and I_j is maximal. ▮

Thus our partition is *the set of maximal* intervals of X.

3.4. PROPERTIES OF INTERVALS

PROPOSITION 3.11

If μ is a cycle in a graph (X, Γ) and $x \in \mu$, then x is its own successor and predecessor. μ is strongly connected.

Proof. Recall that a cycle $\mu[x_1, x_2, \ldots, x_n]$ has $x_1 = x_n$, so that clearly $x_1 \in \hat{\Gamma}x_1$ and $x_1 \in \hat{\Gamma}^{-1}x_i$. Let $x \in \mu$, and consider $\mu' = \mu'[x_1, x]$. Then $\nu = \mu[x_1, \ldots, x, \ldots, x_{n-1}, \mu']$ is such that $x \in \hat{\Gamma}x$ and $x \in \hat{\Gamma}^{-1}x$. ∎

Let I be an interval with head h. I has the following properties:

1. If $x \in I$, $h \leq x$.
2. If I contains any strongly connected subgraph $(S, \Gamma_S) \subset (I, \Gamma_I)$, $h \in S$.
3. If I is not strongly connected but contains a strongly connected subgraph $(S, \Gamma_S) \subsetneq (I, \Gamma_I)$, then there does not exist any $s \in S$ and $i \in I \sim S$ such that $s \geq i$.
4. If $\{\Gamma^{-1}h\} \cap I \neq \varnothing$, then (I, Γ_I) contains a maximal strongly connected subgraph (S, Γ_S) and $\{\Gamma^{-1}h\} \cap I = L \subset S$. The set L is called the set of *back latches* of I.
5. If $L \neq \varnothing$, then $S = I \cap (\bigcup_{l \in L} \{\Gamma^{-k}l\}$, $k = 0, 1, 2, \ldots, n = \#I$.

Proofs of the above properties are trivial. Hence, if an interval I contains any circuits [i.e., if any loops from the program (X, Γ) are in I], their presence can be detected if $\{\Gamma^{-1}h\} \cap I \neq \varnothing$. Each such circuit passes through h.

3.4.1. Points of Confluence

The set $E = \{e \in I | \{\Gamma e\} \sim I \neq \varnothing\}$ comprises the set of *exit vertices* of I; $E_S = \{e \in S | \{\Gamma e\} \sim S \neq \varnothing\}$ is the set of *exit vertices* of S. It is not necessary that $E_S \subset E$.

In the remainder of this chapter we shall observe the following notational conventions:

I	interval
S	strongly connected subinterval of I
h	interval head of I
L	set of back latches of I
E	set of exit nodes of I
E_S	set of exit nodes of S

DEFINITION **3.12**

Let $x \in I$. If for each $\mu = [h, x_1, x_2, \ldots, e_j]$ and for each $e_j \in E$ we have $x \in \mu$, we call x a *point of confluence* of I. Similarly, if $x \in S$ and for each $\mu_s = [h, x_1, \ldots, e_{s_j}]$, where $e_{s_j} \in E_s$, we have $x \in \mu_s$, we call x a point of confluence of S. These sets will be denoted

C set of points of confluence of I

C_S set of points of confluence of S

Again, looking at (X, Γ) as the flow graph of a program, for each interval I, C_S is composed of those basic blocks which must be executed in each program loop in I each time the loop is executed.

PROPOSITION **3.13**

Given a maximal interval I with strongly connected subsets S,

$$C_s = I \bigcap_{e_S \in E_S} (\mathfrak{B}_{e_S} \cup \{e_s\})$$

Proof. If there were only one exit node e_S, it is clear that $C_S = \mathfrak{B}_{e_S} \cup \{e_S\}$ by definition; i.e., every $b \in \mathfrak{B}_{e_S}$ lies on every path $\mu = [h, \ldots, e_S]$, and certainly $C_S \in \mu$.

If there are several exit points $\{e_S\}$, any point of confluence C_S must lie on every path $\mu = [h, \ldots, e_{S_j}]$ and hence must be in $\mathfrak{B}_{e_{S_j}}$. ∎

COROLLARY **3.12.1**

If $S \neq \varnothing$, then $h \in C_S$.

COROLLARY **3.12.2**

If $E_S = \{e\}$, then $e \in C_S$.

Historical note. In early global optimizing compilers, such as those of F. E. Allen and the FORTRAN (H) compiler, the control flow graph was partitioned into nested strongly connected subgraphs and optimization analysis was performed in an inner to outer manner on these subgraphs. As will become obvious in the sequel, this process leaves something to be desired in the determination of information concerning the redundancy of certain subexpressions in basic blocks.

The interval was selected as a more desirable partitioning by John Cocke, since it has the decided advantage of being a partially ordered single-entry subgraph which may contain nested strongly connected subgraphs which are themselves single-entry. These properties are easily exploited in much of the analysis that follows.

4 DERIVED INTERVALS, REDUCIBLE AND IRREDUCIBLE GRAPHS

4.1. DERIVED GRAPHS

Let $\mathscr{I}$ be the unique partition of $G = (X, \Gamma)$ into maximal intervals. Consider the graph $(\mathscr{I}, \Gamma_{\mathscr{I}})$ whose vertices $I \in \mathscr{I}$ are maximal intervals of G and where we define $\Gamma_{\mathscr{I}}$ as

$$I_j \in \Gamma_{\mathscr{I}} I_i \text{ iff for } x_k \in I_i \text{ and } x_l \in I_j \text{ we have } x_l \in \Gamma x_k \text{ where } i \neq j$$

It is possible to partition $(\mathscr{I}, \Gamma_{\mathscr{I}})$ into a set $\mathscr{I}^{(2)}$ of intervals through application of Algorithm 3.8. Similarly, a graph $(\mathscr{I}^{(2)}, \Gamma_{\mathscr{I}^{(2)}})$ consisting of vertices $I^{(2)} \in \mathscr{I}^{(2)}$ may be partitioned into a set $\mathscr{I}^{(3)}$ of intervals $I^{(3)}$, etc.

We define $(\mathscr{I}^{(n)}, \Gamma_{\mathscr{I}^{(n)}})$ to be the nth derived graph of (X, Γ).

Since $G = (X, \Gamma)$ is always finite for our purposes, it is clear that for a given graph G there exists an integer $N = N(G)$ such that

$$(\mathscr{I}^{(N)}, \Gamma_{\mathscr{I}^{(N)}}) = (\mathscr{I}^{(N+1)}, \Gamma_{\mathscr{I}^{(N+1)}}) = \cdots$$

There are two cases of interest: (1) $\mathscr{I}^{(N)}$ is a single interval, or (2) $\mathscr{I}^{(N)}$ comprises several intervals. In the former case, G is said to be *reducible*, and in the latter, *irreducible* (of order N).

4.2. AN IRREDUCIBLE GRAPH

We shall now give an example of a graph and its partitioning into intervals. Vertices will be denoted by a circled integer which will serve to identify the vertex. In general we shall not draw arrowheads on the arcs connecting vertices, but shall follow the

CONVENTION

Arcs leaving a vertex will exit from the bottom of the circle.
Arcs entering a vertex will enter the top of the circle.

Hence in Fig. 4.1. we have the following arcs: (1, 2), (1, 3), (3, 2), and (2, 3).

Consider the graph in Fig. 4.2. The intervals in Fig. 4.2 are

$$I_1: \{1, 2, 3, 4, 5\}$$
$$I_2: \{6, 7, 8, 9, 16\}$$
$$I_3: \{10, 11, 12, 13, 14, 15\}$$

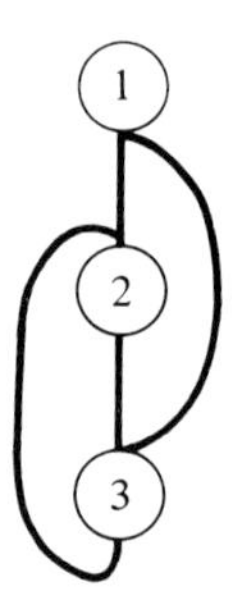

Figure 4.1

Vertex 6 cannot be in I_1, since $\Gamma^{-1}(6) = \{4, 5, 9, 15\}$ and $\{9, 15\} \not\subset I_1$. Similarly, vertex 10 cannot be in I_2 since $\Gamma^{-1}(10) = \{9, 15, 3\}$ and $3 \in I_1$.

The derived graph $(\mathscr{I}, \Gamma_{\mathscr{I}})$ is shown in Fig. 4.3. In this case $(\mathscr{I}, \Gamma_{\mathscr{I}})$ is irreducible, since if I_1 is taken as the interval head for $I_1^{(2)}$, I_2 cannot be included in $I_1^{(2)}$ since $I_3 \in \Gamma^{-1}I_2$ and $I_3 \notin I_1^{(2)}$. Similarly, I_3 cannot be placed in $I_1^{(2)}$. Now if I_2 becomes the interval head for $I_2^{(2)}$, I_3 cannot be included since $I_1 \in \Gamma^{-1}I_3$ and $I_3 \notin I_2^{(2)}$. Hence $(\mathscr{I}^{(2)}, \Gamma_{\mathscr{I}^{(2)}}) = (\mathscr{I}, \Gamma_{\mathscr{I}})$ and $(\mathscr{I}, \Gamma_{\mathscr{I}})$ is irreducible.

4.3. A REDUCIBLE GRAPH

In Fig. 4.4 we show the reduction of a completely reducible graph.

4.4. FACTS ABOUT IRREDUCIBLE GRAPHS

We shall now prove a significant fact about irreducible graphs.

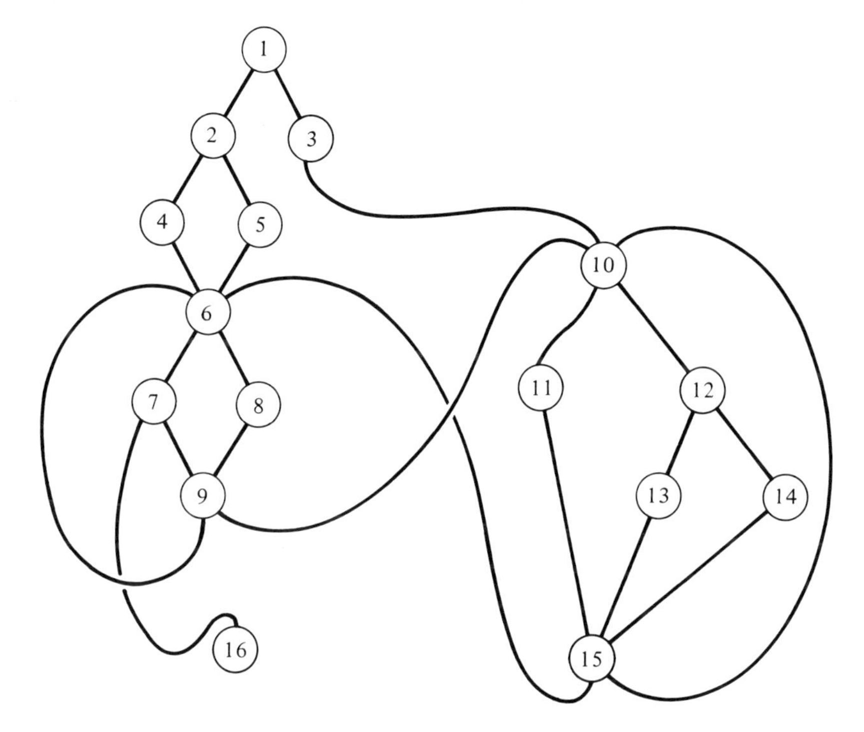

Figure 4.2

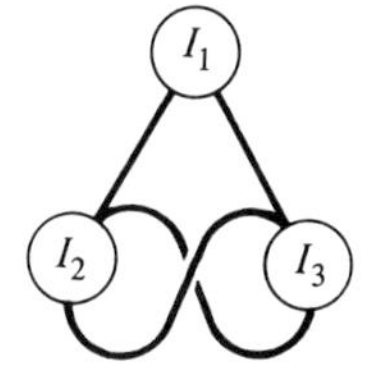

Figure 4.3

THEOREM **4.1**

If $G = (X, \Gamma)$ is irreducible of order N, then either

1. There exists $x \in X$ such that $x \notin \hat{\Gamma} e$, and G is not connected, or

2. G contains a circuit with multiple entry points, i.e., a circuit μ with at least two vertices $x, y \in \mu$ having immediate predecessors in $X \sim \mu$.

Proof. If $G^{(N)} = (\mathcal{I}^{(N)}, \Gamma_{\mathcal{I}^{(N)}})$ is the Nth derived graph of $G = (X, \Gamma)$, it is clear that any circuit of $G^{(N)}$ is a circuit of G. Similarly, if $x \in \hat{\Gamma} e$ in G, then $x^{(N)} \in \hat{\Gamma}_{\mathcal{I}^{(N)}} e^{(N)}$, in $G^{(N)}$, where $x^{(N)}$ is the node of $G^{(N)}$ containing x, and $e^{(N)}$ is the entry node of $G^{(N)}$. Hence proving Theorem 4.1 for $G^{(N)}$ is sufficient.

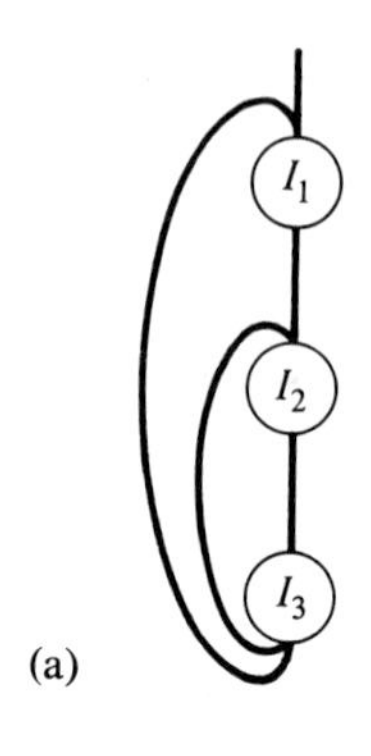

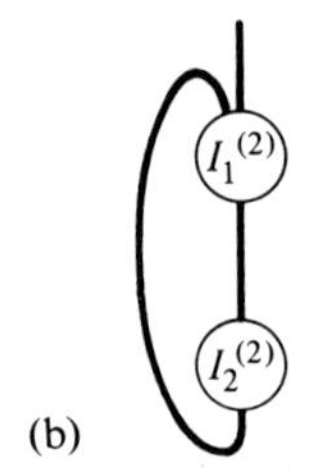

(c) $I_1^{(3)}$

Figure 4.4 (a) $(\mathcal{I}, \Gamma_{\mathcal{I}})$, $\#\mathcal{I} = 3$; (b) $(\mathcal{I}^{(2)}, \Gamma_{\mathcal{I}^{(2)}})$, $\#\mathcal{I}^{(2)} = 2$; (c) $(\mathcal{I}^{(3)}, \Gamma_{\mathcal{I}^{(3)}})$, $\#\mathcal{I}^{(3)} = 1$.

Given that condition 1 does not hold, we assume that $x^{(N)} \in \hat{\Gamma}_{\mathcal{I}^{(N)}} e^{(N)}$ for all $x^{(N)} \in \mathcal{I}^{(N)}$. Now, with the exception of $e^{(N)}$, each $\#\Gamma^{-1}_{\mathcal{I}^{(N)}}(x^{(N)}) \geq 2$, for if not, $x^{(N)}$ could be included in the same interval as its immediate predecessor.

Thus, since each $x^{(N)} \neq e^{(N)}$ has at least two predecessors, each such $x^{(N)}$ has at least one predecessor distinct from $e^{(N)}$. The existence of a cycle is dependent, therefore, on the following proposition: If each of n vertices in a graph K has at least one predecessor from among the remaining $n - 1$ vertices, then there exists a cycle μ of K with $l(\mu) \geq 2$. We shall prove this by induction:

The proposition is clearly true for $n = 2$ or $n = 3$; hence we assume that the proposition is true for all graphs consisting of at most $k - 1$ vertices, and we let K be a graph with k vertices. Pick a subgraph L of K consisting of $k - 1$ vertices. If each of these $k - 1$ vertices has a predecessor from among the remaining $k - 2$ vertices, the induction hypothesis guarantees the existence of a cycle involving at least two of the $k - 1$ vertices. If not, then there is at least one vertex of L having as its only immediate predecessor the vertex $v \in K \sim L$, while $\varnothing \neq \Gamma^{-1} v \subseteq H$. If $\Gamma v \cap \Gamma^{-1} v \neq \varnothing$, we are through. Otherwise, since L is cycle-free, some path of length at most $\#(L \sim \Gamma v)$ from a vertex in Γv must intersect $\Gamma^{-1}v$; i.e., there is a cycle in K involving v.

Returning to $G^{(N)}$, let $H^{(N)} = [x^{(N)} = y_1^{(N)}, \ldots, y_k^{(N)}]$ be a minimal strongly connected subgraph of $G^{(N)}$ containing $x^{(N)}$. We now claim that there exist at least two vertices $y_i^{(N)}, y_j^{(N)} \in H^{(N)}$ such that each predecessor is in $\sim H^{(N)}$. To see this, construct two subsets of $H^{(N)}$, viz.,

$$A = \{a \in H^{(N)} | \{\Gamma^{-1}a\} \subset \{H^{(N)}\}\}$$
$$B = H^{(N)} \sim A$$

LEMMA **4.1.1**

$$\#B \geq 2.$$

Proof. Assume not. $B \neq \varnothing$, for then $H^{(N)}$ would be disconnected from $G^{(N)}$. Hence, $B = \{b\}$ and b would be the circuit head of $\mu = H^{(N)}$, since $H^{(N)}$ is a minimal strongly connected subregion of $G^{(N)}$. But then μ could be collapsed to a single interval with interval head b, a contradiction to the irreducibility of $G^{(N)}$. Q.E.D. lemma

Hence $\#B \geq 2$, and $G^{(N)}$ contains a multientry circuit. ∎

Note that by the remark following Algorithm 3.8, condition 1 of Theorem 4.1 is not possible.

Note that not every vertex in $G^{(N)}$ need have more than one predecessor from outside the *maximal* strongly connected subgraph, as is shown in Fig. 4.5 with node 4.

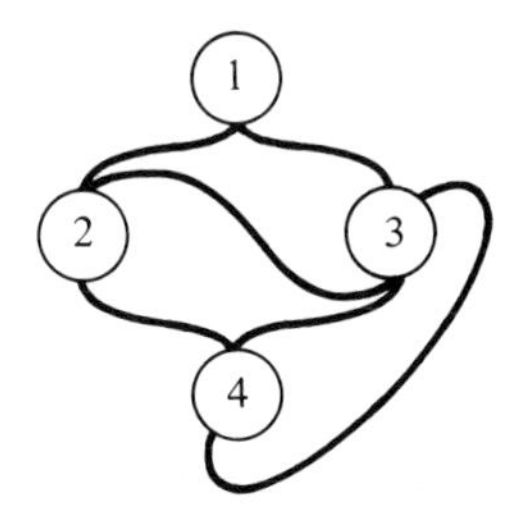

Figure 4.5

We may extend Theorem 4.1 slightly to obtain

THEOREM **4.2**

In the irreducible graph $G^{(N)}$ every prime circuit is multientry.

Proof. If there were a single-entry prime circuit μ, we could form an interval I headed by the vertex $x \in \mu$ which has immediate predecessors outside of μ and at least one of the successors of x. ∎

Note that Theorem 4.2 concludes only that the prime circuits of $G^{(N)}$ are multientry. For example, in Fig. 4.6 the graph is clearly irreducible, yet the circuit [2, 3, 4] is single-entry, while the prime circuits [3, 4], [2, 3], and [2, 4] are multientry.

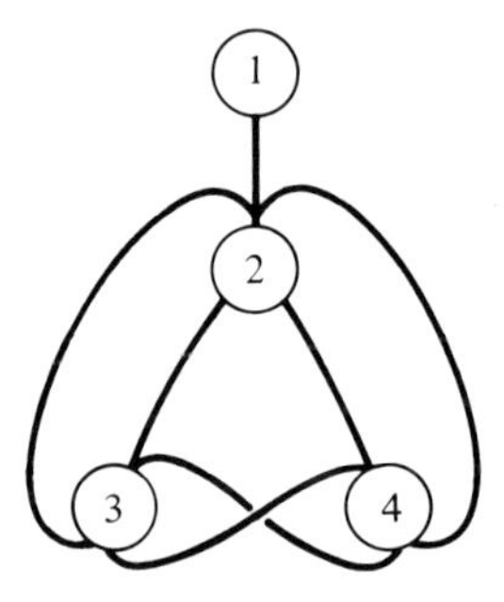

Figure 4.6

THEOREM **4.3**

In the irreducible graph $G^{(N)}$ every vertex of every prime circuit μ has an immediate predecessor outside of μ.

Proof. Every vertex in $G^{(N)}$ except $e^{(N)}$ has at least two predecessors. If any vertex of a prime circuit μ had two predecessors from within the circuit, μ would contain a proper subcircuit and, hence, not be prime. ∎

5 VERTEX ORDERING ALGORITHMS

5.1. MOTIVATION

We shall now consider the problem of locating the strongly connected subgraphs of an interval. While this task is relatively simple for people looking at neatly drawn graphs, we must take into account the fact that these graphs will be represented within the one-dimensional store of a computer. It is therefore necessary to clarify the concept that, in an interval, the strongly connected subgraph lies "above" all the back-latches. This may be achieved by imposing yet another order relation on the nodes of each interval. We shall follow an adaptation of an ordering developed by C. P. Earnest of Computer Sciences Corporation.

In the following algorithms we shall use the notational conventions from the previous chapters. We recall then, that

1. I is some interval.
2. h is the head of I.
3. $L = \{l \in I \mid h \in \Gamma l\}$, the set of back-latches of I.

The algorithms below will assign to each vertex $x \in I$ an integer $\Theta(x)$. Alternatively, we shall write x_k, where $\Theta(x_k) = k$, with k an integer. We shall strive for an ordering, then, which satisfies the following criteria:

1. If $y \in \Gamma x \sim \{h\}$, then $\Theta(x) < \Theta(y)$.
2. If $l \in L$, then $\{x \in I \mid \Theta(h) \leq \Theta(x) \leq \Theta(l)\}$ is a strongly connected subset of I.
3. $\Theta(h) < \Theta(x)$ for every $x \in I(h) \sim \{h\}$.

5.2. THE BASIC NUMBERING ALGORITHM

ALGORITHM **5.1** (*Basic Numbering Algorithm*)
For an interval I with head h,

1. Let $S_1 = \{h\}$; set $\Theta(h) = 1$.
2. Let $p = 1$, $k = 1$.
3. If $n_p \in S_k$ and there exists $x \in I \cap (\Gamma n_p \sim S_k)$, then
   a. For each y such that $p < \Theta(y) \leq \# S_k$, increase $\Theta(y)$ by 1.
   b. Define $\Theta(x) = p + 1$, $S_{k+1} = S_k \cup \{x\}$.
   c. Increase k and p each by 1.
   d. Return to step 3.
4. If $p > 1$, decrease p by 1 and return to step 3; otherwise stop.

Example 5.2

The interval in Fig. 5.1 is transformed by the Basic Numbering Algorithm to the order shown in Fig. 5.2. For simplicity, we shall identify nodes by circled letters.

We note that the ordering in Fig. 5.2 is but one possible ordering that could be assigned the graph in Fig. 5.1. Indeed, step 3 of Algorithm 5.1 is applied to each $x \in \Gamma n_p$ such that $x \notin S$, and if there are several such x, different orderings will result.

There are several qualities of Fig. 5.1 that are lacking in Fig. 5.2. In the former it is readily apparent where the strongly connected subintervals are and which vertices are in each, whereas this quality is not so obvious in Fig. 5.2.

5.3. THE STRICT NUMBERING ALGORITHM

We shall resolve this difficulty in Algorithm 5.3. We shall first state that a set $A = \{a_1, a_2, \ldots\}$ of vertices ordered by Algorithm 5.1 has the *natural ordering* if for each $a_i, a_j \in A$, $\Theta(a_i) < \Theta(a_j)$ whenever $i < j$. A path $T = [a_i, a_{i_2}, \ldots, a_{i_n}]$ is called a *forward track* iff $\Theta(a_{i_j}) < \Theta(a_{i_k})$ for $j < k$.

We shall have need of a notation for the manipulation of certain ordered

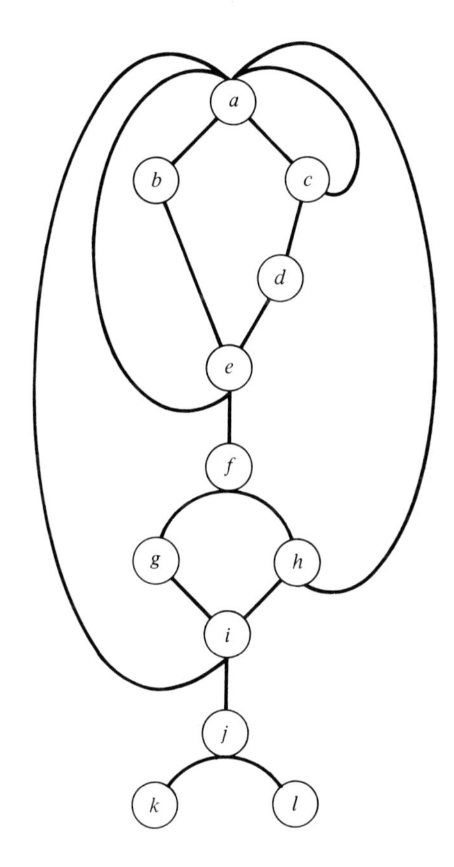

Figure 5.1 Interval before the Basic Numbering Algorithm.

sets. In particular, it will be necessary to append one set to another in an order-preserving manner akin to direct sums.

Define for two sets A, B with $A \cap B = \varnothing$ the sum

$$C = A \oplus B = \{a_1, a_2, \ldots, a_{\#A}, b_1, b_2, \ldots, b_{\#B}\}$$

If A or B is not ordered, we require only that the elements of A and B be listed separately.

Similarly, if A is some ordered set and B an arbitrary set, we define the difference

$$C = A \ominus B = \{a_i \in A \sim B \mid a_i \text{ precedes } a_j \text{ in } C \text{ iff } a_i \text{ precedes } a_j \text{ in } A\}.$$

We now state

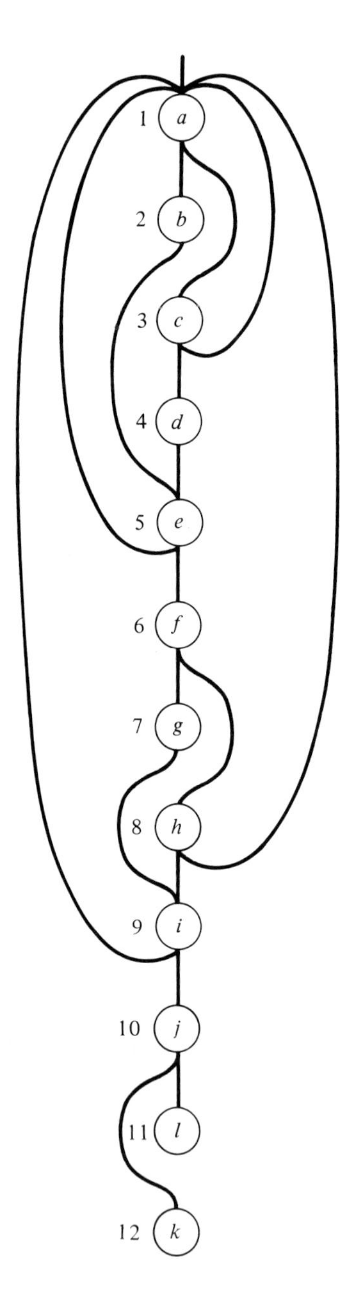

Figure 5.2 Interval after one possible ordering by the Basic Numbering Algorithm.

ALGORITHM **5.3** (*Strict Numbering Algorithm*)

Let I be an interval with head h, and let $L = \{l \in I | h \in \Gamma l\}$. Then to order the nodes of I,

1. If $L = \varnothing$, apply Algorithm 5.1 and stop.
2. Define $\Theta(h) = 1$, $v = 2$, $M_1 = \varnothing$.
3. To each $l_i \in L$ let there correspond sets $S_{i1} = \{l_i\}$ and $\Lambda_{i1} = \{l_i\}$.
4. For each l_i define $S_i = \{h\} \oplus \lim_j S_{ij}$, where

$$S_{i,j+1} = \Lambda_{i,j+1} \oplus (S_{ij} \ominus \Lambda_{i,j+1})$$
$$\Lambda_{i,j+1} = \Gamma^{-1}\Lambda_{ij} \sim \{h\}$$

5. Form a sequence $S_{i_1}, S_{i_2}, \ldots, S_{i\#L}$ from the S_i so that

$$\#S_{i_k} \leq \#S_{i_{k+1}}, \qquad k = 1, 2, \ldots, \#L - 1$$

6. For $k = 1, 2, \ldots, \#L$, number the nodes of $S_{i_k} \ominus M_k$ consecutively from v, let $M_{k+1} = M_k \cup S_{i_k}$, and increment v by $\#(S_{i_k} \sim M_k)$.
7. Let $S = \{h\} \cup_k S_{i_k}$. If $I = S$, stop.
8. Let $N_1 = S$, $t = 1$.
9. If $I = N_t$, stop. Otherwise, if there exists a node $x \in I \cap (\Gamma^{\$}S \sim N_t)$, then
   a. For each y such that $v \leq \Theta(y) \leq \#N_t$, increase $\Theta(y)$ by 1.
   b. Define $\Theta(x) = v$; let $p = v$, $N_{t+1} = N_t \cup \{x\}$.
   c. Increase t by 1.
10. If $n_p \in N_t$ and there exists $z \in I \cap (\Gamma n_p \sim N_t)$, then
    a. For each w such that $p < \Theta(w) \leq \#N_t$, increase $\Theta(w)$ by 1.
    b. Define $\Theta(x) = p + 1$, $N_{t+1} = N_t \cup \{z\}$.
    c. Increase t and p each by 1.
    d. Return to step 10.
11. If $p > \#S + 1$, decrease p by 1 and return to step 10; otherwise return to step 9.

The result of applying the strict numbering algorithm to the graph of Fig. 5.2 is shown in Fig. 5.3. It is now quite simple to identify all the vertices in each strongly connected subinterval of the original interval simply by identifying a backward branch l and including all the vertices between h and l.

5.4. INTERESTING PROPERTIES OF THE ALGORITHMS

We shall now note several properties of Algorithms 5.1 and 5.3, which will be referred to as BNA and SNA, respectively.

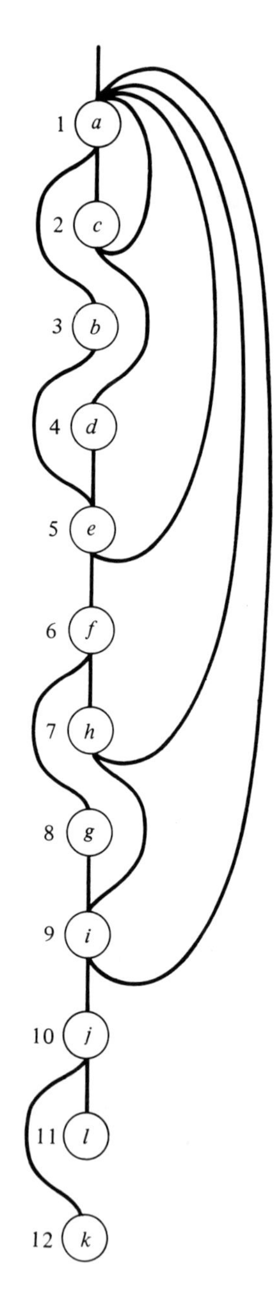

Figure 5.3 Interval after the Strict Numbering Algorithm.

PROPERTY 5.4

After application of the BNA, every circuit includes a strongly connected region; i.e., if $h \in \Gamma l$, for $l \in L$, either $l = h$ or there is a path $\mu = [h, \ldots, l]$ such that for every $x \in \mu$, $1 < \Theta(x) \leq \Theta(l)$.

Proof. If $l \neq h$, then at the time l was inserted into S_k, such a path existed from h to l. Later insertions into $S_{k'}$ cannot affect the existence of this path, and if the index of any vertex in this path was subsequently increased, so was the index of each of its successors, including l. ∎

PROPERTY 5.5

After application of the BNA, for each $l \in L$, there exists a forward path from h to every other node within the backward branch. That is, if $h \in \Gamma l$, then for every $x \in I$, $1 < \Theta(x) \leq \Theta(l)$, there exists a path $\mu = [h, x_1, \ldots, x]$ with $\Theta(x_i) < \Theta(x_j)$ for $i < j$.

Proof. When x was inserted into S_k, there was a direct path to x from h; thus at that time the property was clearly true. For any vertex inserted later, the relative ordering of the vertices in this path remained unchanged, although the indices of some of the vertices in the path may have been incremented. ∎

PROPERTY 5.6

After application of the BNA, there is a forward path from h to every other vertex in I.

Proof. Obvious. ∎

THEOREM 5.7

If $a, b \in I$, $a \neq b$, and $a \leq b$, then after application of the BNA, $\Theta(a) < \Theta(b)$.

Proof. From the definition of back dominance, every path from h to b includes all back dominators of b. Thus the direct path from h to b at the time b was inserted into S_k must have included a, and so $\Theta(a) < \Theta(b)$. ∎

PROPERTY 5.8

In the SNA, the sets S_i are strongly connected subsets of I.

Proof. We need show that for $x, y \in S_i$ there exists a path $\mu = [x, \ldots, y] \subseteq S_i$.

Clearly, there exist paths $\mu_1 = [h, \ldots, y] \subseteq S_i$ and $\mu_2 = [x, \ldots, l_i] \subseteq S_i$. Since $h \in \Gamma l_i$, the path $\mu_2\mu_1 = [x, \ldots, y]$ suffices. ∎

PROPERTY 5.9

Let $S_i = \{h = s_{i_1}, s_{i_2}, \ldots, s_{i_n} = l_i\}$ be listed in an order produced by step 4 of the SNA. Then $s_{i_k} \in S_i \sim \{h\}$; if there exists $s_{i_j} \in S_i$ such that

$s_{i_k} \in \Gamma s_{i_j}$, we have $j < k$. A fortiori, for any simple path $\mu = [x_{i_1}, x_{i_2}, \ldots, x_{i_m}] \in S_i \sim \{h\}$, $\Theta(x_{i_j}) < \Theta(x_{i_k})$ for $j < k$.

Proof. Let $s_{i_j} \in S_i$, $s_{i_k} \in S_i \sim \{h\}$, $s_{i_k} \in \Gamma s_{i_j}$. Assume that $s_{i_k} \in \Lambda_{i\nu}$, and hence $s_{i_k} \in S_{i\nu}$ for some ν in the construction of S_i in step 6. Then $s_{i_j} \in \Lambda_{i,\nu+1}$ and therefore $s_{i_j} \notin S_{i\nu} \ominus \Lambda_{i,\nu+1}$, but $s_{i_j} \in \Lambda_{i,\nu+1} \oplus (S_{i\nu} \oplus \Lambda_{i,\nu+1})$. Thus in S_i we must have s_{i_j} preceding s_{i_k}.

Now, let S_a be some other strongly connected set produced in step 4 which precedes S_i in the sequence of sets defined in step 5, and furthermore let S_a be the first set in this ordering which contains s_{i_j}. s_{i_j} can be in no set S_b which precedes S_a in this ordering, for since $s_{i_j} \in \Gamma^{-1} s_{i_k}$, we would have $s_{i_j} \in S_b$, a contradiction. Then $\Theta(s_{i_j}) < \Theta(s_{i_k})$. The truth of the assertion follows immediately. ∎

PROPERTY **5.10**

Let $x_1, x_2 \in I \sim S$, $x_2 \in \Gamma x_1$. Then $\Theta(x_1) < \Theta(x_2)$. Furthermore, if $s \in S$, $z \in I \sim S$, then $\Theta(s) < \Theta(z)$.

Proof. Steps 9, 10, and 11 order the nodes of $I \sim S$ in a manner equivalent to the BNA. The results follow directly. ∎

PROPERTY **5.10**

The SNA does not destroy forward tracks. That is, if there is a forward track from a to b after the BNA, then there is a forward track from each a to b after the SNA.

Proof. Follows directly from Properties 5.9 and 5.10. ∎

In the sequel we shall revert to the notational convention of identifying a vertex's name and order number by assuming that the SNA has already been applied to the interval which we are discussing.

It will be necessary at times to refer to the formal limit of a sequence. In this discussion we shall always be using finite sets, so that any sequence of subsets of elements from the set X must necessarily contain a finite number of distinct terms, in fact no more than $2^{\#X}$.

We let $\{\alpha_\nu\}$, $\nu = 1, 2, \ldots$, be a sequence of subsets of X. We say that

$$\lim \alpha_\nu = \lim_{\nu \to \infty} \alpha_\nu$$

exists and write

$$\lim \alpha_\nu = \alpha$$

If, from some point N on, $\alpha_N = \alpha_{n+1} = \cdots$, we have $\alpha = \alpha_N$.

EXERCISE

If G is a reducible graph, produce an algorithm which imposes the strict order on all the nodes of G. What must be done if G is irreducible?

6 LATTICE ALGEBRA AND THE REDUCTION OF IRREDUCIBLE GRAPHS

6.1. A LATTICE ALGEBRA

We shall have need for the lattice algebra defined by the following set of postulates. Let S be a system with binary operations addition $(+)$ and multiplication $(\cdot)$ defined so that for arbitrary $a, b, c \in S$ each of the following holds:

1. *Closure:*

$$a \cdot b \text{ is a unique element of } S$$
$$a + b \text{ is a unique element of } S$$

2. *Commutativity:*

$$a \cdot b = b \cdot a$$
$$a + b = b + a$$

3. *Associativity:*

$$a \cdot (b \cdot c) = (a \cdot b) \cdot c$$
$$a + (b + c) = (a + b) + c$$

4. *Distributive law:*

$$a \cdot (b + c) = (a \cdot b) + (a \cdot c)$$
$$(a + b) \cdot c = (a \cdot c) + (b \cdot c)$$

5. *Absorption:*

$$a \cdot (a + b) = a$$
$$a + (a \cdot b) = a$$

6. Let $\mathcal{B} = S \cup \{1, 0\}$, $+$ and $\cdot$ as defined above as unique elements of $\mathcal{B}$ with the additional requirements that

$$a + 0 = a$$
$$1 \cdot a = a$$

Then $\mathcal{B}$ forms a distributive lattice algebra. In general we shall write $a \cdot b$ as ab.

6.1.1. Some Simple Properties of the Algebra

We note the following important propositions.

PROPOSITION **6.1**

$aa = a.$

Proof.

$$\begin{aligned} aa &= a(a + ab) \qquad &(5)\dagger \\ &= a &(5) \end{aligned}$$ ∎

PROPOSITION **6.2**

$a + a = a.$

Proof.

$$\begin{aligned} a + a &= a + aa \qquad &(1) \\ &= a &(5 \text{ with } b = a) \end{aligned}$$ ∎

PROPOSITION **6.3**

$ab = a. \quad a + b = b.$

Proof.

$$\begin{aligned} a + b &= ab + b \qquad &(\text{hypothesis}) \\ &= b + ab &(2) \\ &= b + ba &(2) \\ &= b &(5) \end{aligned}$$ ∎

PROPOSITION **6.4**

$a + b = b \implies ab = a.$

Proof.

$$\begin{aligned} ab &= a(a + b) \\ &= a \end{aligned}$$ ∎

PROPOSITION **6.5**

$0 \cdot a = 0.$

†Refers to the properties defined above.

Proof.

$$\begin{aligned} 0 \cdot a &= 0(a + 0) && (6) \\ &= 0(0 + a) && (2) \\ &= 0 && (5) \end{aligned}$$ ∎

PROPOSITION **6.6**

$a + 1 = 1.$

Proof. $1 \cdot a = a \implies 1 + a = 1 \quad (3)$ ∎

THEOREM **6.7**

$a + (b \cdot c) = (a + b) \cdot (a + c).$

Proof.

$$\begin{aligned} (a + b) \cdot (a + c) &= ((a + b) \cdot a) + ((a + b) \cdot c) && (4) \\ &= (a + (a + b) \cdot c) && (5) \\ &= a + ((a \cdot c) + (b \cdot c)) && (4) \\ &= (a + (a \cdot c)) + (b \cdot c) && (3) \\ &= a + (b \cdot c) && (5) \end{aligned}$$ ∎

We shall assign $\cdot$ a higher binding than $+$ so that $a \cdot b + c$ means $(a \cdot b) + c$. It can be shown that $+$ and $\cdot$, 0 and 1 are dual symbols so that the dual of any theorem may be derived by simply interchainging symbols.

Alternative symbols for $+$ and $\cdot$ are $\cup$ and $\cap$ or $\vee$ and $\wedge$. We prefer the former since they permit more lucid exposition of results. Compare $a + bc = (a + b)(a + c)$ to $a \cup (b \cap c) = (a \cup b) \cap (a \cup c)$ or $a \vee (b \wedge c) = (a \vee b) \wedge (a \vee c)$.

6.2. NODE SPLITTING

We shall now develop an application of this distributive lattice algebra to the reduction of irreducible graphs.

6.2.1. An Heuristic Approach

Let $G = (X, \Gamma)$ be irreducible, so that its only possible partitioning into intervals is discrete. Then, with the exception of the entry node, every node in G has at least two predecessors.

It is possible to duplicate certain nodes in such a graph in such a manner that the resulting graph is further reducible and equivalent to it in the sense that any path in the original graph exists in the transformed graph.

An example should make this clear. Because of the complexity of the graph involved, it will be necessary to draw arcs with arrowheads rather than by the convention established previously.

Example 6.8

The graph G in Fig. 6.1(a) is irreducible, while the equivalent graph G' in Fig. 6.1(b) can be reduced to the form shown in Fig. 6.1(c), which is irreducible. In turn, $\mathscr{g}$ can be transformed to $\mathscr{g}'$ [Fig. 6.1(d)], which is completely reducible.

This transformation is referred to as *node splitting*. Node splitting is not a unique transformation. Indeed, in addition to the symmetric transformation where nodes b and c of Fig. 6.1(b) are interchanged (generating nodes b' and b'' in place of c' and c''), the graph G'' in Fig. 6.2 is further reducible.

6.2.2. An Analysis of our Approach

It is clear by inspection of the graphs in Example 6.8 that the node splitting was accomplished by duplicating at least one node from each cycle in the graph. It is more interesting to note that one node was chosen from each cycle to head each constructed interval of the transformed graph (except, of course, the entry node a).

Now if $G = (X, \Gamma)$ is not reducible, G must be decomposable into $\#X$ intervals. To reduce G so that $G' = (X', \Gamma)$ is further reducible, there must occur a coalescing of nodes of G so that G' reduces to no more than $\#X - 1$ intervals. Since the entry node will head one of these intervals, no more than $\#X - 2$ nodes may be chosen from the cycles of G to head intervals of G'. In general there will be more than $\#X - 2$ cycles in G so that some selection techniques must be employed in order to minimize the actual number of nodes selected. For example, the graph G of Example 6.8 has four cycles, viz., $[b, c]$, $[b, d]$, $[c, d]$, and $[b, c, d]$.

6.2.3. The Permanent

A simple method of selecting nodes can be derived from the distributive lattice algebra constructed earlier. If we give the logical values of *and* and *or* to $\cdot$ and $+$, respectively, then consider the equation

$$\pi = (b + c)(b + d)(c + d)(b + c + d)$$

where, for exposition purposes, b, c, and d correspond to the nodes in the example. Here we intend to select one of the vertices b or c, one of b or d, one of c or d, and one of b, c, or d from each cycle. Expanding the equation, we have

$$\begin{aligned}\pi &= (b + c)(b + d)(c + d)(b + c + d)\\ &= (b + cd)(c + d)(b + c + d)\\ &= (b + cd)(c + d)\\ &= bc + cd + db\end{aligned}$$

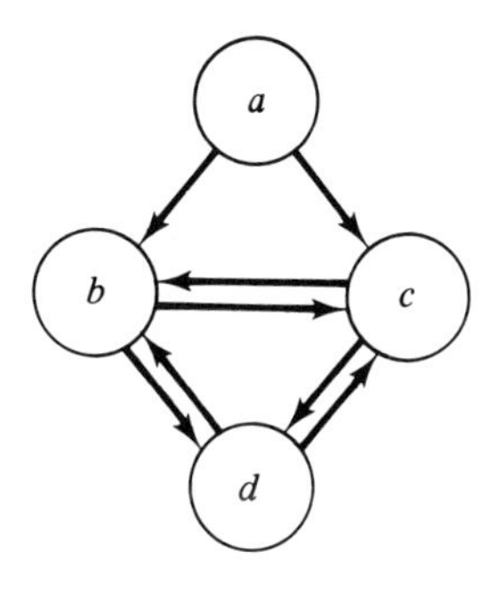

(a)
Irreducible Graph G

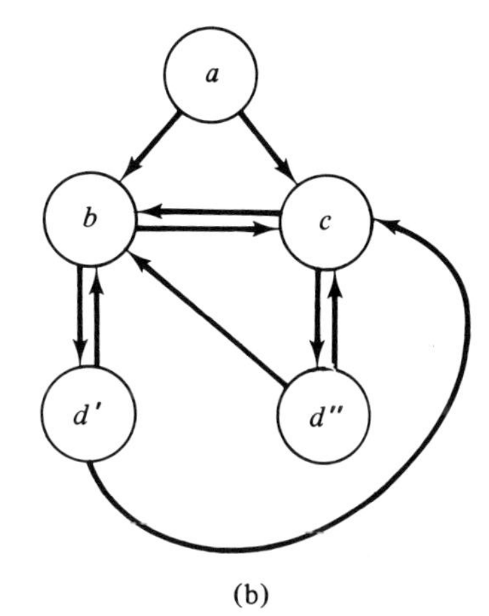

(b)
Equivalent Reducible Graph G'

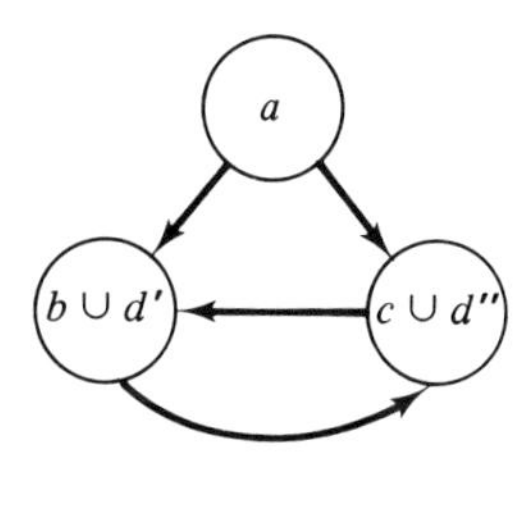

(c)
Irreducible Graph $\mathcal{I}$ Derived From G'

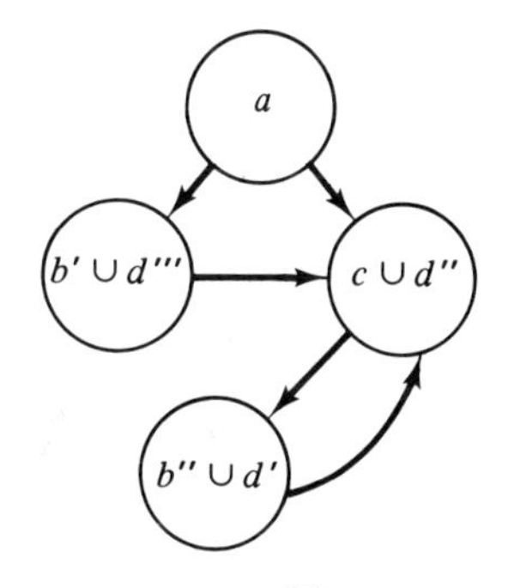

(d)
Equivalent Completely Reducible Graph G'

Figure 6.1

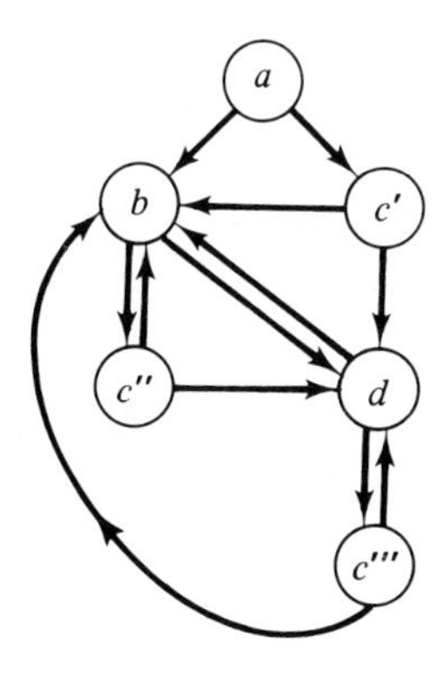

Figure 6.2 The graph G'' is further reducible and equivalent to the graph G in Fig. 6.1(a).

so that π is equal to *b and c, or c and d, or b and d.* In Example 6.8 these are precisely the nodes selected to head new intervals.

It should be noted that the cycle $[c, d] \subset [b, c, d]$, while, by absorption $(c + d)(b + c + d) = (c + d)$. Recall that a *prime cycle* is a cycle which contains no proper subcycles; it is clear that a product such as π is unaffected by the deletion of nonprime cycles. Techniques for identifying prime cycles are discussed in Chapter 7.

DEFINITION **6.9**

To each vertex x_i of a graph $G = (X, \Gamma)$ let there correspond a Boolean variable x^i. To each (prime) cycle μ_j let there correspond the sum

$$C^j = \sum_{x_i \in \mu_j} x^i$$

Then, letting μ_j range over all the (prime) cycles of G, we define π, the *permanent* of G, as

$$\pi = \prod_j C^j = \prod_j \sum_{x_i \in \mu_j} x^i$$

where $+$ and $\cdot$ are the operations of the distributive lattice algebra (1–6).

6.2.3.1. Properties of the Permanent

THEOREM **6.10**

Let π be the permanent of G. If π is reduced such that it is the sum of products, then in any summand each cycle of G is represented by at least one factor. Equivalently, if $S_k = n^1 \cdot n^2 \cdot \cdots \cdot n^l$ is a summand of π in reduced form, the deletion of each n_i and its associated arcs is sufficient to render G cycle-free.

Proof. π is a product of sums of Boolean variables, each of which corresponds to a vertex in some cycle of G, and all prime cycles of G are represented in the product. Boolean variables are variables assuming either 0 or 1 as a value, and so by the earlier results if any one summand is equal to 1, π will equal 1. A summand will equal 1 only if each factor is 1.

Let $S_k = n^1 \cdot n^2 \cdot \cdots \cdot n^l$ be a summand of π in reduced form, and without loss of generality assume that $\mu = [a_1, \ldots, a_j]$ is a prime cycle of G such that no $n_i \in \mu$. Then

$$(a^1 + a^2 + \cdots + a^j)(n^1 \cdot n^2 \cdot \cdots \cdot n^l) = a^1 n^1 n^2 \cdots n^l + \cdots + a^j n^1 n^2 \cdots n^l$$

and there must be precisely j distinct summands with $l + 1$ distinct factors. But $\pi \cdot (a^1 + a^2 + \cdots + a^j) = \pi$ since

$$(a^1 + a^2 + \cdots + a^j)^2 = (a^1 + a^2 + \cdots + a^j)$$

to give precisely the summands of π; however, this implies that there was a summand in π which when multiplied by some $a^i \in \mu$, yields S_k, a contradiction. ∎

THEOREM **6.11**

Let $S_k = n^1 \cdot n^2 \cdot \cdots \cdot n^l$ be a summand of π. Then S_k is minimal in the sense that G will not be cycle-free unless each node of G corresponding to a factor of S_k is deleted.

Proof. π is assumed to be in reduced form, so that, e.g., the absorption law has been applied. Hence it is not the case that for any products α, β, both α and $\alpha\beta$ are summands of π.

Therefore assume that S_k is not minimal. Then there is a node, say n_1, which need not be deleted from G in order to render G cycle-free. Hence n^1 is redundant in S_k. Since to each cycle of G there corresponds at least one factor in S_k and n^1 is redundant, there must be some other node, say n_2, such that n_1 and n_2 are in the same prime cycle.

Without loss of generality assume that n_1 is the only redundant factor of S_k (otherwise the argument is repeated), and let $n^1 = 0$, $n^2 = \cdots = n^l = 1$, and set all other variables to 0. Then π is of the form

$$\begin{aligned}\pi &= (\cdots n^1 + \cdots + n^2 + \cdots)(\cdots + n^3 + \cdots) \cdots (\cdots + n^j + \cdots) \\ &\quad \cdots (\cdots + n^l + \cdots) \\ &= (\cdots 0 + 1) \cdot (1) \cdot \cdots \cdot (1) \cdot \cdots \cdot (1) = 1\end{aligned}$$

Now, since π is in reduced form, no two summands contain precisely the same factors, and since $n^2, \ldots, n^l$ are all in S_k and $n^2 \cdot \cdots \cdot n^l$ is not a factor by itself, each summand contains some factor equal to 0 and is therefore equal to 0 by Proposition 6.5. Thus $\pi = 0 + 0 + \cdots + S_k + \cdots + 0 = 0$, since $S_k = 0 \cdot 1 \cdot 1 \cdot \cdots \cdot 1$, a contradiction. ∎

6.2.3.2. Applicability of the Permanent

It is clear that no summand of π can have more factors than the number of distinct prime cycles of G. Since e, the entry node of G, is not a member of any cycle, we have yet to show that there are not more than $\#(X \sim \{e\}) - 1 = \#X - 2$ prime cycles of G if we intend to further reduce G.

This is certainly not possible if each vertex of $X \sim \{e\}$ is itself a loop, for there would then be $\#X - 1$ prime cycles. However, loops can all be removed by identifying each vertex as an interval and thereby reducing G without changing the cardinality of G'.

Hence it can be assumed that every cycle of G contains at least two distinct vertices.

THEOREM **6.12**

If each prime cycle of G contains at least two vertices, no summand of π is composed of more than $\#X - 2$ factors. Equivalently, if π has a summand with $\#X - 1$ factors, then each prime cycle of G is a loop.

Proof. If π had a summand with $\#X - 1$ factors, then π could have no other summand, for any other such summand would have fewer factors and since each factor of the other summand would be a factor of the larger summand, the larger summand would be absorbed into the summand with fewer factors. Hence π is a monomial.

Assume that each prime cycle of G contains at least two vertices. Arbitrarily pick any vertex n_i of some prime cycle, and set $n^i = 0$ while $n^j = 1$ for all $j \neq i$. Then, as in the above proofs, we have each of the factors of π equal to 1 so that, on the one hand, $\pi = 1$, while, on the other hand, if π were the monomial consisting of $\#X - 1$ factors, we would have $\pi = 0$, the desired contradiction. ∎

COROLLARY **6.13**

Any graph of n vertices can be disconnected by deleting no more than $n - 1$ vertices.

Corollary 6.13 is intuitively obvious and, because of the results of Theorems 6.10 and 6.11, could have been used to prove Theorem 6.12 in one line. We shall give the algebraic proof because of the interesting contrast between the distributive lattice algebra and the more usual commutative associative algebra over a field.

Theorem 6.12 guarantees us that the evaluation of the permanent π of G will yield at least one set of nodes to head intervals in the equivalent *split graph* so that reduction may continue on any nontrivial irreducible graph.

PROPOSITION **6.14**

Let $n^1, \ldots, n^k$ be k distinct variables in a distributive lattice algebra. Then

$$\prod_{1 \leq i < j \leq k} (n^i + n^j) = \sum_{i=1}^{k} \prod_{\substack{j \neq i \\ 1 \leq j \leq k}} n^j = n^2 n^3 \cdots n^k + n^1 n^3 \cdots n^k + \cdots + n^1 n^2 \cdots n^{k-1}$$

Proof. By absorption,

$$\prod_{1 < j \leq k} (n^1 + n^j) = n^1 + n^2 n^3 \cdots n^k = n^1 + \prod_{j=2} n^j$$

and in general

$$\prod_{c < j \leq k} (n^c + n^j) = n^c + \prod_{j=c+1}^{k} n^j$$

so that

$$\prod_{1 \le i < j \le k} (n^i + n^j) = (n^1 + \prod_{j=2}^{k} n^j)(n^2 + \prod_{j=3}^{k} n^j) \cdots (n^c + \prod_{j=c+1}^{k} n^j) \cdots (n^{k-1} + n^k)$$

$$= \prod_{i=1}^{k} (n^i + \prod_{j=i+1}^{k} n^j)$$

We next combine terms from the left, observing that

$$(n^1 + \prod_{j=2}^{k} n^j)(n^2 + \prod_{j=3}^{k} n^j) = n^1 n^2 + n^1 \prod_{j=3}^{k} n^j + \prod_{j=2}^{k} n^j (n^2 + \prod_{j=3}^{k} n^j)$$

$$= n^1 n^2 + n^1 \prod_{j=3}^{k} n^j + \prod_{j=2}^{k} n^j$$

$$(n^1 + \prod_{j=2}^{k} n^1)(n^2 + \prod_{j=3}^{k} n^j)(n^3 + \prod_{j=4}^{k} n^j)$$

$$= (n^1 n^2 + n^1 \prod_{j=3}^{k} n^j + \prod_{j=2}^{k} n^j)(n^3 + \prod_{j=4}^{k} n^j)$$

$$= n^1 n^2 n^3 + n^1 n^2 \prod_{j=4}^{k} n^j + n^1 \prod_{j=3}^{k} n^j + \prod_{j=2}^{k} n^j$$

or, in general,

$$\prod_{i=1}^{c} (n^i + \prod_{j=i+1}^{k} n^j) = \prod_{j=1}^{c} n^j + \sum_{l=1}^{c} \prod_{\substack{j \ne l \\ 1 \le j \le k}} n^j$$

The proposition follows immediately by letting $c = k - 1$ and multiplying by $(n^{k-1} + n^k)$. ▮

COROLLARY **6.15**

The complete graph may be rendered cycle-free only by the deletion of all but one vertex.

Proof. Follows immediately from Proposition 6.14. ▮

6.3. A NODE SPLITTING ALGORITHM

6.3.1. More Heuristics

We now have the machinery with which to determine the interval heads for the split graph, and so we shall develop an algorithm for establishing the placement of the split nodes. Since we favor an intuitive approach, we shall make liberal use of examples.

Example 6.16

Consider the graph of Fig. 6.1(c). We compute the permanent

$$\pi = (B + C)$$

where $B = \{b, d'\}$ and $C = \{c, d''\}$. Hence either B or C could be used as the interval head for the split graph. We choose C, and onto C we attach each prime cycle involving C, duplicating vertices as necessary. The transformation is shown for reference in Fig. 6.3. We delete the arc connecting C with the original B since B is not going to head an interval and therefore can have no predecessors which are also successors.

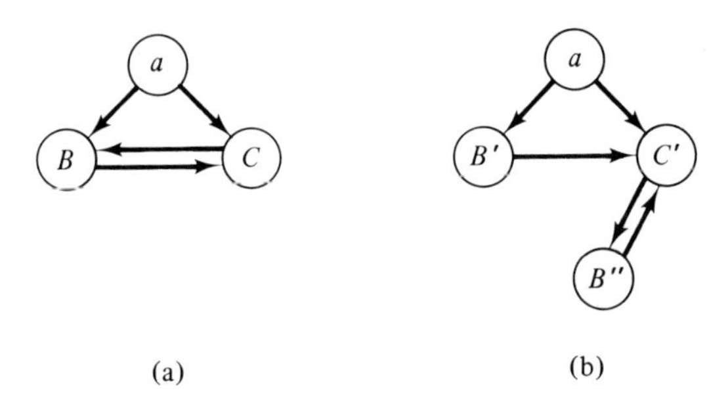

Figure 6.3

Example 6.17

Consider the graph of Fig. 6.4(a). Its permanent is

$$\pi = (b + c)(c + d)(b + d) = bc + cd + bd$$

We select the factors of any summand, say b and c, as interval heads. To them we connect the prime cycles in which each is involved, getting the partial graph shown in Fig. 6.4(b).

Note that the prime cycle $[b, c]$ is duplicated and the prime cycle $[c, d]$ does not exist in the interval headed by b, while $[b, d]$ is not in the interval headed by c.

We first delete one of the redundant prime cycles $[b, c]$, noting that this can be accomplished by connecting either interval head with the other, as shown in Fig. 6.4(c).

Next we note that d' must connect to b and that d'' must connect to c to establish the connectivity of the strongly connected subgraph of G. Finally, we add a to G' and connect a with the interval heads b and c. The result is shown in Fig. 6.4(d).

Suppose instead that we chose $\{c, d\}$ from π. We would again attach to each of c and d the prime cycles in which they are involved [Fig. 6.4(e)]. It is convenient to identify the d heading an interval with the d in the cycle $\{c, d\}$ in the interval headed by c.

We next note the presence of two $[c, d]$ cycles and arbitrarily delete the one headed by d; no modification of connectivity is necessitated by this deletion [Fig. 6.4(f)].

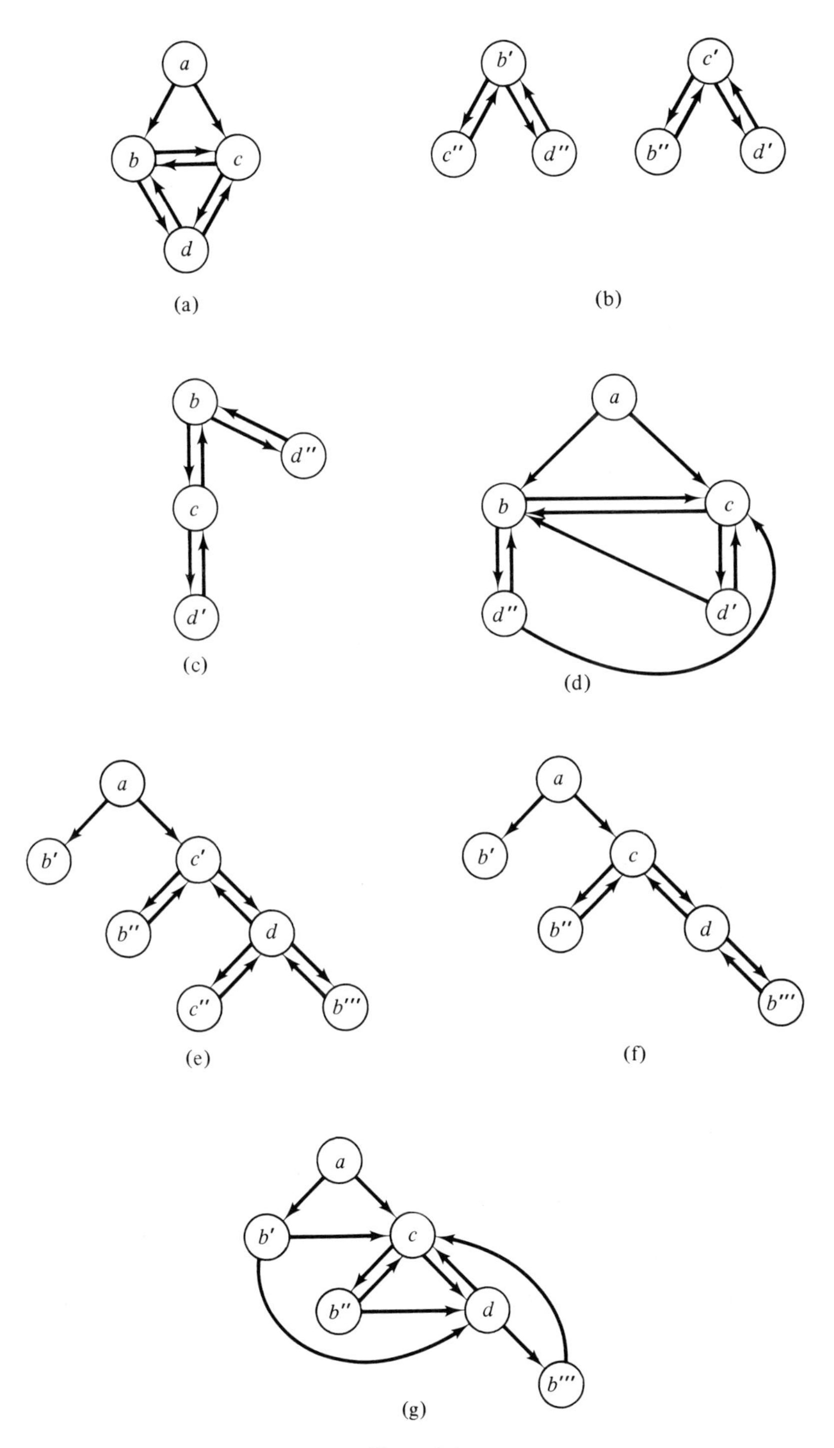

Figure 6.4

The connectivity of the original graph is next extablished: b' must connect to c'; b' must also connect to d, which can be done since d heads an interval. B'' must connect with d, and b''' must connect with c, again possible since c heads an interval. The resulting split graph is shown in Fig. 6.4(g).

6.3.2. Graph Equivalence

Let the permanent of G be written as

$$\pi = \sum \prod_{j=j(i)} n^j = \sum_i S_i$$

Then for $S_l = \prod_{j=j(l)} n^j$ we define $F_l = \{n_j \in X \mid n^j \text{ is a factor of } S_l\}$.

As indicated in Examples 6.16 and 6.17, $G' = (X', \Gamma')$, the split graph of $G = (X, \Gamma)$, is equivalent in flow to G, while there is a one-to-many correspondance between the elements of X and X'. This correspondance is determined by some multivalued function $\varphi: X \rightarrow X'$, where φ^{-1} is a single-valued function. For each $n_i \in X$ there corresponds a maximal set $N_i = \{n_i^q \in X' \mid \varphi^{-1}(n_i^q) = n_i\}$ such that there is no set M, with $N_i \subset M \subseteq X'$, where $\varphi(M) = \{n_k\}$.

DEFINITION **6.18**

Two graphs $G = (X, \Gamma)$ and $G' = (X', \Gamma')$ are *equivalent* if there is a multivalued function

$$\varphi: X \longrightarrow 2^{X'}$$
$$n_i \longmapsto N_i$$

such that φ^{-1} is single-valued with $\varphi^{-1}N_i = n_i$ and $N_i \subseteq \Gamma' N_j$ iff $n_i \in \Gamma n_j$.

6.3.3. The Algorithm

In Algorithm 6.19 we shall use the letters *a–h* as constants; the letters *i–z* are variables. We recall that in a cycle $\mu = [x_1, x_2, \ldots, x_p]$ where $x_i \in \Gamma x_{i-1}$ for $i = 2, 3, \ldots, p$, $x_1 \in \Gamma x_p$, and $l(\mu) = p$.

The entry node e of G will be represented by n_1.

ALGORITHM **6.19**

Assume that $G = (X, \Gamma)$ is a directed graph that cannot be further reduced into intervals. To construct an equivalent reducible split graph $G' = (X', \Gamma')$,

1. From $\pi = \sum_k \prod_{j=j(k)} n^j$ select an arbitrary summand S_a. a will remain constant through this algorithm. Form $F_a = \{n_j \in X \mid n^j \text{ is a factor of } S_a\}$.

2. To each $n_j \in F_a \cup \{n_1\}$ let there correspond a unique element n_j^j, a set $\hat{N}_j^1 = \{n_j^j\}$, and a number $\nu(j) = 1$.

3. Construct a set Y of symbols n_p^q:

$$Y = \{n_p^q \mid n_p \in X \sim (F_a \cup \{n_1\}) \text{ and } n_q \in F_a \cup \{n_1\}\}$$

4. For each set $\hat{N}_m^{\nu(m)} = \{n_m^m = n_{m_1}^m, n_{m_2}^m, \ldots, n_{m\nu(m)}^m\}$, let $p = 1$, and
   a. If $p \leq \nu(m)$, then define $(\Gamma' n_{m_p}^m)^0 = \varnothing$ and a number $\gamma(m_p, m) = 0$; then
   b. For each $n_i \in \Gamma n_{m_p} \subseteq X$ either
      (1) If $n_i \in F_a$, define $(\Gamma' n_{m_p}^m)^{\gamma(m_p, m)+1} = (\Gamma' n_{m_p}^m)^{\gamma(m_p, m)} \cup \{n_i^m\}$, and set $\gamma^{(m_p, m)} \longleftarrow \gamma^{(m_p, m)+1}$, or
      (2) Define $\hat{N}_m^{\nu(m)+1} = \hat{N}_m^{\nu(m)} \cup \{n_i^m\}$, define

$$(\Gamma' n_{m_p}^m)^{\gamma(m_p, m)+1} = (\Gamma' n_{m_p}^m)^{\gamma(m_p, m)} \cup \{n_i^m\},$$

      and set $\nu(m) \longleftarrow \nu(m) + 1$, $\gamma(m_p, m) \longleftarrow \gamma(m_p, m) + 1$.
   c. Set $p \longleftarrow p + 1$.

5. Define $\hat{N}_m = \hat{N}_m^{\nu(m)}$, $\Gamma' n_p^q = (\Gamma' n_p^q)^{\gamma(p,q)}$, $X' = \bigcup_m \hat{N}_m$.

Then $G' = (X', \Gamma')$ is reducible with intervals $\hat{N}_m$ and is equivalent to G.

Proof. We defined a multivalued function

$$\varphi: X \longleftarrow 2^{X'}$$
$$n_j \longmapsto \{n_j^m \in \hat{N}_m\} = N_j$$

where

$$\varphi^{-1}: X' \longrightarrow 2^X$$
$$n_j^m \longmapsto n_j$$

is clearly single-valued.

By construction $n_p^q \in \Gamma n_r^s$ if $n_p \in \Gamma n_r$, so that $N_p \subseteq \Gamma' N_q$. Applying φ^{-1} to both sides of $N_p \subseteq \Gamma' N_q$ yields $n_p \in \Gamma n_q$ so that G' is equivalent to G.

To establish the reducibility of G', it is sufficient to show that $\hat{N}_m \sim \{n_m^m\}$ is cycle-free since step 2 places $n_m^m \in \hat{N}_m$ and for any $n_j^m \in \hat{N}_m$, $n_j^m \in \hat{\Gamma}' n_m^m$ by step 4.b(2). However, this is clear since for any cycle μ of G, F_a contains some $n_k^m \in N_m$.

The maximality of $\hat{N}_m$ follows by construction. Finally, since $\#F_a \leq \#X - 2$, $\#\{n_k^k \in X'\} \leq \#X - 1$. ∎

6.3.4. An Application of the Algorithm

Example 6.20

Consider the graph G in Fig. 6.5. G is clearly irreducible and has prime cycles $\mu_1 = [b, c, d, e]$, $\mu_2 = [f, c]$, $\mu_3 = [g, d, e]$, $\mu_4 = [e, h]$, and $\mu_5 = [b, f, g]$.

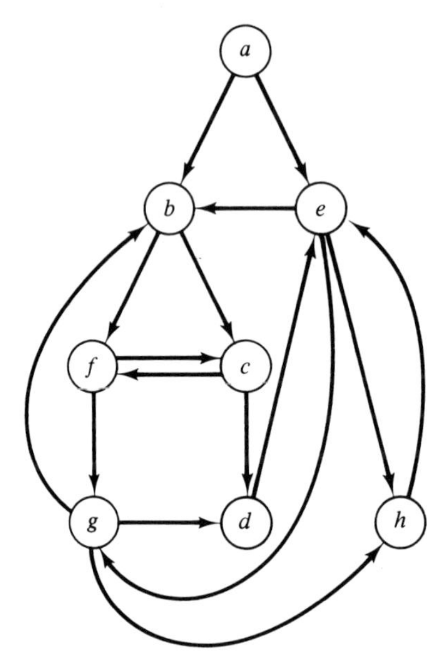

Figure 6.5

We have

$$\begin{aligned}\pi &= (b + c + d + e)(f + c)(g + d + e)(e + h)(b + f + g)\\ &= bfgh + bce + bcdh + ceg + chg + dfh + ef\end{aligned}$$

Let

$$F = \{e, f\}, \qquad \text{so } \hat{E}^0 = \{e^2\}, \quad \hat{F}^0 = \{f^3\}, \quad \hat{A}^0 = \{a^1\}$$

Then we obtain

$$\hat{A} = \{a^1, b^1, c^1, d\}$$

with

$$\begin{aligned}\Gamma' a^1 &= \{b^1, e^2\}\\ \Gamma' b^1 &= \{c^1, f^3\}\\ \Gamma' e^1 &= \{d^1, f^3\}\\ \Gamma' d^1 &= \{e^2\}\\ \hat{E} &= \{e^2, b^2, c^2, d^2, g^2, h^2\}\end{aligned}$$

with

$$\begin{aligned}\Gamma' e^2 &= \{b^2, g^2, h^2\}\\ \Gamma' b^2 &= \{c^2, f^3\}\end{aligned}$$

$$\Gamma' c^2 = \{d^2, f^3\}$$
$$\Gamma' d^2 = \{e^2\}$$
$$\Gamma' g^2 = \{b^2, d^2, h^2\}$$
$$\Gamma' h^2 = \{e^2\}$$
$$\hat{F} = \{f^3, b^3, c^3, d^3, g^3, h^3\}$$

with

$$\Gamma' f^3 = \{c^3, g^3\}$$
$$\Gamma' b^3 = \{c^3, f^3\}$$
$$\Gamma' c^3 = \{d^3, f^3\}$$
$$\Gamma' d^3 = \{e^2\}$$
$$\Gamma' g^3 = \{b^3, d^3, h^3\}$$
$$\Gamma' h^3 = \{e^2\}$$
$$X' = \{a^1, b^1, b^2, b^3, c^1, c^2, c^3, d^1, d^2, d^3, e^2, f^2, g^1, g^2, g^3, h^2, h^3\}$$

$G' = (X', \Gamma')$ is shown in Fig. 6.6.

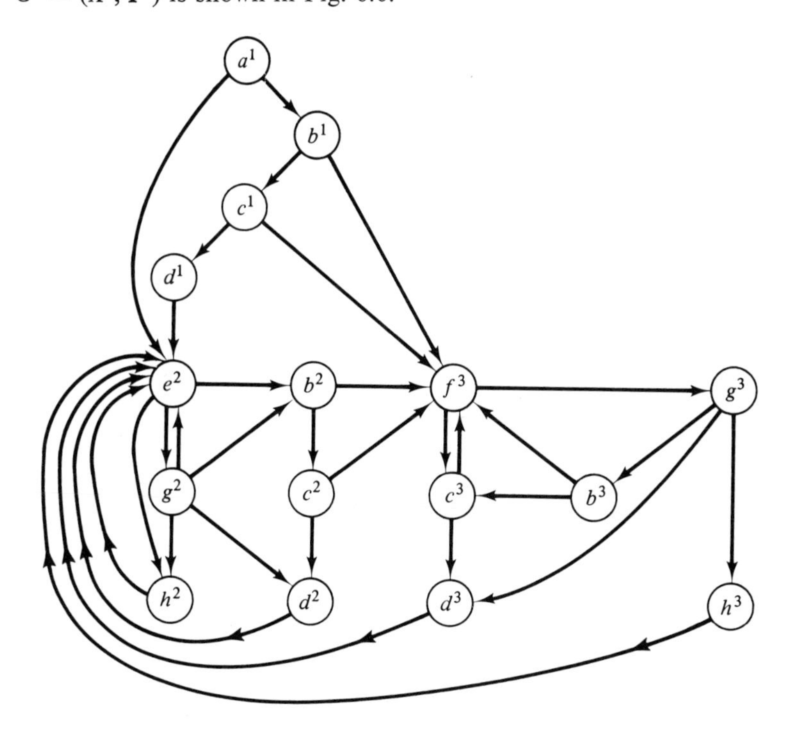

Figure 6.6

Example 6.21

Using the graph of Fig. 6.5, let $F = \{b, f, g, h\}$; therefore $\hat{A}^0 = \{a^1\}$, $\hat{B}^0 = \{b^2\}$, $\hat{F}^0 = \{f^3\}$, $\hat{G}^0 = \{g^4\}$, and $\hat{H}^0 = \{h^5\}$.

We then obtain

$$\hat{A} = \{a^1, e^1\}$$

with

$$\begin{aligned} \Gamma' a^1 &= \{b^2, e^1\} \\ \Gamma' e^1 &= \{g^4, h^5\} \\ \hat{B} &= \{b^2, c^2, d^2, e^2\} \end{aligned}$$

with

$$\begin{aligned} \Gamma' b^2 &= \{c^2, f^3\} \\ \Gamma' c^2 &= \{d^2, f^3\} \\ \Gamma' d^2 &= \{e^2\} \\ \Gamma' e^2 &= \{g^4, h^5\} \\ \hat{F} &= \{f^3, c^3, d^3, e^3\} \end{aligned}$$

with

$$\begin{aligned} \Gamma' f^3 &= \{c^3, g^4\} \\ \Gamma' c^3 &= \{f^3, d^3\} \\ \Gamma' d^3 &= \{e^3\} \\ \Gamma' e^3 &= \{g^4, h^5\} \\ \hat{G} &= \{g^4, d^4, e^4\} \end{aligned}$$

with

$$\begin{aligned} \Gamma' g^4 &= \{b^2, d^4, h^5\} \\ \Gamma' d^4 &= \{e^4\} \\ \Gamma' e^4 &= \{g^4, h^5\} \\ \hat{H} &= \{h^5, e^5\} \end{aligned}$$

with

$$\begin{aligned} \Gamma' h^5 &= \{e^5\} \\ \Gamma' e^5 &= \{g^4, h^5\} \end{aligned}$$

See Fig. 6.7.

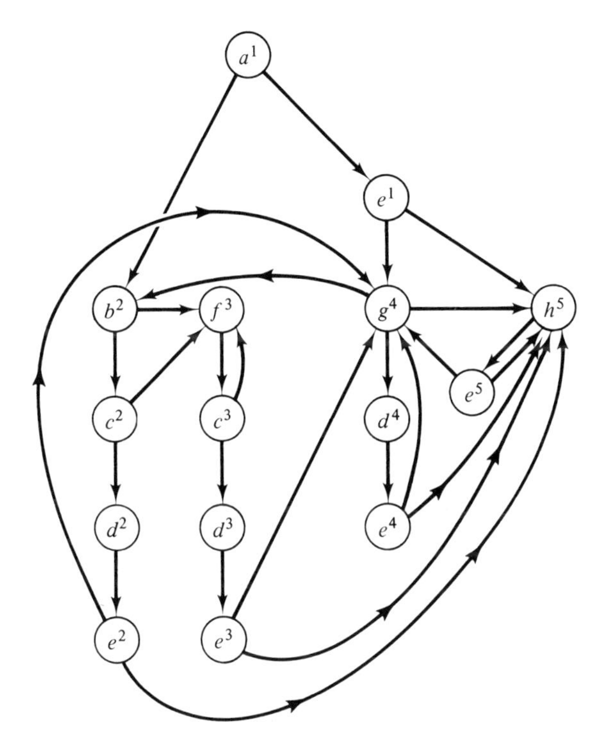

Figure 6.7

6.4. AN ALTERNATIVE METHOD OF NODE SPLITTING

We shall mention an alternative method of node splitting derived by John Cocke. The method requires less analysis but generally involves a greater number of transformations than Algorithm 6.19.

Let $G' = (X', \Gamma')$ be the irreducible limit graph of G, $\#X' > 1$, and let $\mathfrak{B}$ be the maximal chain of back-dominating nodes of G'; i.e., $\mathfrak{B} = \{b_0, b_1, \ldots, b_k\}$ satisfies the following properties:

1. b_0 is the derived interval containing the entry node of G.
2. $b_j \leq b_l$ for $0 \leq j < l \leq k$.
3. There exists some $b \in X' \sim \beta$ such that $b_k \leq b$.
4. β is maximal with respect to 1, 2 and 3.

Then select any node $p \in \hat{\Gamma}' b_k$ such that the minimal distance $\delta(b_k, p)$ from b_k to p is equal to $\max_{t \in \hat{\Gamma}' b_k} \delta(b_k, t)$. (See Chapter 7.)

Corresponding to p, we introduce a new node p^* and modify the connectivity of G' by

1. $\mathbf{\Gamma}'p^* = \mathbf{\Gamma}'p$.

2. Pick some $t \in \mathbf{\Gamma}'^{-1}p$ such that $\delta(b_k, t) = \delta(b_k, p) - 1$. Define $\mathbf{\Gamma}'^{-1}p^* = \{t\}$, $\mathbf{\Gamma}'^{-1}p = \{\mathbf{\Gamma}'^{-1}p\} \sim \{t\}$.

Then the modified graph has at least one interval fewer than the unmodified graph (prove!).

Example 6.22

Consider the graph of Fig. 6.8(a). The back dominator chain $\mathfrak{B} = \{a, b\}$. Selecting $g \in \hat{\mathbf{\Gamma}}d$, we obtain the graph in Fig. 6.8(b), which reduces to the graph of Fig. 6.8(c). Selecting H, we obtain Fig. 6.8(d), which reduces to Fig. 6.8(e). The sequence continues through Fig. 6.8(f), (g), (h), (i), (j), (k), and (l).

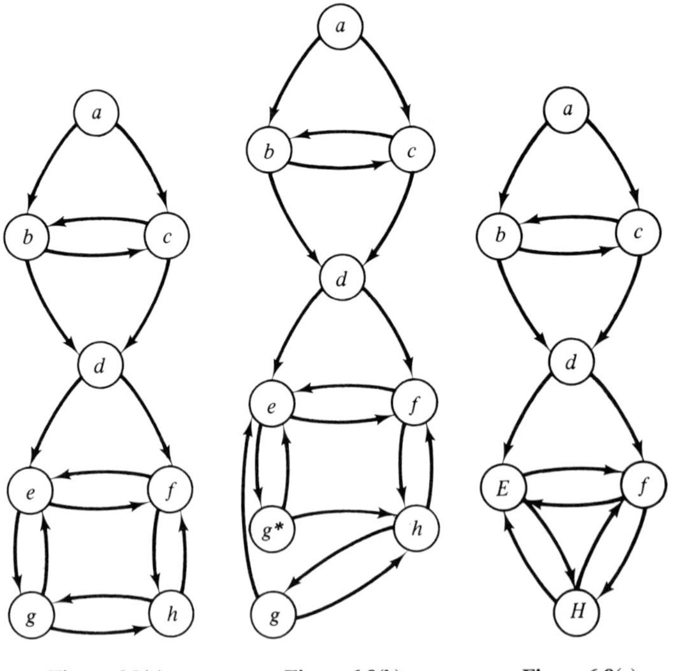

Figure 6.8(a) Figure 6.8(b) Figure 6.8(c)

EXERCISE

Combine Algorithm 6.19 with the method of Cocke to produce a reduction algorithm that requires possibly fewer transformations. When are the methods equivalent?

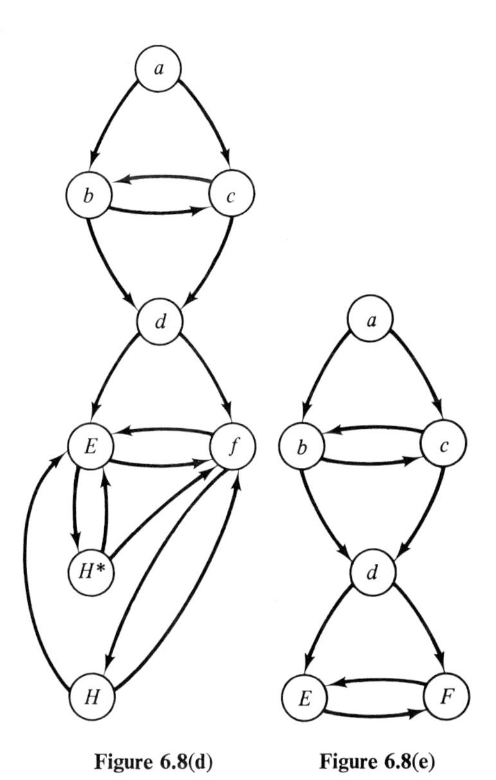

Figure 6.8(d) **Figure 6.8(e)**

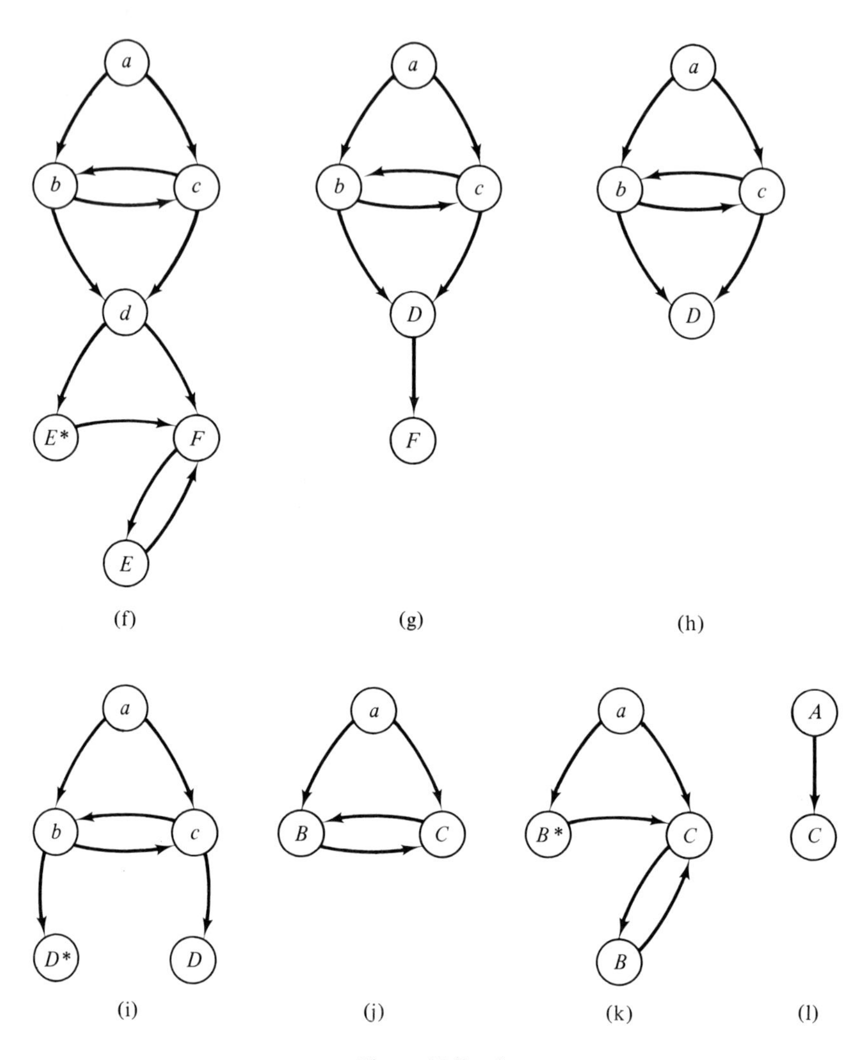

Figure 6.8(f to l)

7 THE CONNECTIVITY MATRIX AND PRIME CYCLES

7.1. THE CONNECTIVITY MATRIX

Let $G = (X, \Gamma)$ be a finite directed graph with $\# X = n$. Thus far we have considered Γ to be a function,

$$\Gamma : X \longrightarrow 2^X$$

Order the elements of X in some manner and consider the n vectors

$$\begin{aligned}
\bar{x}_1 &= (1, 0, 0, \dots, 0)\\
\bar{x}_2 &= (0, 1, 0, \dots, 0)\\
\bar{x}_3 &= (0, 0, 1, \dots, 0)\\
&\vdots\\
\bar{x}_n &= (0, 0, 0, \dots, 1)
\end{aligned}$$

where each $\bar{x}_i$ corresponds uniquely to some $x_i \in X$ and their componentwise sums

$$y_j = \sum_{i=1}^{n} b_{ij}\bar{x}_i$$

where each $b_{ij} = 0$ or 1. Clearly, there are 2^n distinct vectors y_j, and these may be interpreted as *characteristic vectors* or *characteristic functions* on 2^X via the obvious isomorphism $y_j = \sum_{i=1}^{n} b_{ij}\bar{x}_i \longleftrightarrow Y_j \in 2^X$, where $Y_j = \{x_i \in X \mid y_{jk} = 1, k = 1, \dots, n\}$. For notational simplicity we shall identify $\bar{x}_i = y_j$ for $i = j$, $j = 1, 2, \dots, n$.

Similarly, we may define an $n \times n$ matrix $C = C_\Gamma = (c_{ij})$ as

$$c_{ij} = \begin{cases} 1 & \text{if} \quad x_j \in \Gamma x_i \\ 0 & \text{if} \quad x_j \notin \Gamma x_i \end{cases}$$

C is the *connectivity matrix* (or *incidence matrix*) of G.

We shall denote the ith row of a matrix C as C_i and the jth column by $C_{,j}$. The matrix $C^T = (c^T_{ij})$, where $c^T_{ij} = c_{ji}$, is called the *transpose matrix of* C. The identity matrix $I = I_n$ is the matrix where $i_{jj} = 1$, $i_{jk} = 0$ for $j \neq k$.

Let $x_i \in X$ and consider the vector $y_j = \bar{x}_i C$, where $\bar{x}_i$ is the corresponding characteristic vector on 2^X and the regular matrix product is taken, i.e.,

$$y_{jk} = \sum_{l=1}^{n} \bar{x}_{il} c_{lk}$$

Then y_j is the characteristic vector of Γx_i. To see this, note that only one component of $\bar{x}_i$ is nonzero, namely $\bar{x}_{ii}$, so that each $y_{jk} = \bar{x}_{ii} c_{ik}$, i.e., $y_i = C_i$.

Now, let $y = \{\cup_{\alpha \in A} x_\alpha \mid A \subset \{1, 2, \ldots, \#X\}, x_\alpha \in X\}$ be some element of 2^X. Then the characteristic vector $\bar{y}$ of y may be written as

$$\bar{y} = \sum_{\alpha \in A} \bar{x}_\alpha$$

and, clearly, $\bar{z} = \bar{y}C = \sum_{\alpha \in A} \bar{x}_\alpha C$. $\bar{z}$ is obviously equal to the sum $\sum_{\alpha \in A} C_\alpha$ of those rows of C corresponding to the 1s of $\bar{y}$, i.e., the sum of those rows of C corresponding to the elements of y. The component $\bar{z}_k$ will be 0 if and only if each $C_{\alpha k} = 0$, $\alpha \in A$. If $\bar{z}_k \neq 0$, then $\bar{z}_k$ must be equal to the number of instances where $C_{\alpha k} = 1$. Hence $\bar{z}_k$ is equal to the number of arcs from y to x_k. If the matrix product were considered as taking place in the distributive lattice algebra described in Chapter 6, $\bar{z}_k$ would be 1 if and only if some $C_{\alpha k} = 1$ and would be 0 otherwise; i.e., in that case $\bar{z}$ would be the characteristic vector of $z = \Gamma y$. As we shall see, both arithmetic interpretations will be of interest, and we shall denote the lattice algebra operation with a radical sign; i.e., the lattice algebra product of a vector $\bar{y}$ and a matrix C will be written as

$$\bar{z} = (\bar{y}C)\surd$$

7.2. POWERS OF THE CONNECTIVITY MATRIX

Since I is the characteristic matrix of the nodes of X (in the order of C), we note that $IC = C$, and

$$(IC)C = I(CC) = IC^2 = C^2$$

Now, the effect of the left-hand side of the above equation is to map the set of immediate successors of each node $x_i \in X$ into its immediate successors; i.e., $C_i^2\surd$ is the characteristic vector of $\Gamma^2 x_i$, while C_{ij}^2 is equal to the number of paths of length 2 from x_i to x_j.

By induction we obtain

THEOREM **7.1**

Let C be the connectivity matrix of the graph $G = (X, \Gamma)$. Then

$$C^k\surd = (CC^{k-1})\surd$$

is the characteristic matrix of Γ^k, in the sense that $c_{ij}^k\surd = 1$ if and only if there exists at least one path of length k from x_i to x_j, i.e., if and only if $x_j \in \Gamma^k x_i$.

The elements c_{ij}^k of C^k represent the number of paths of length k from x_i to x_j.

Denote the componentwise product of two matrices A, B by $A \times B$, i.e.,

$$(A \times B)_{ij} = a_{ij}b_{ij}$$

Then it immediately follows that

PROPOSITION **7.2**

Let H_1 and H_2 be subgraphs of $G = (X, \Gamma)$, and let the $n \times n$ matrices C_{H_1}, C_{H_2}, and C be their respective connectivity matrices. Then the connectivity matrix for $H_1 \triangle H_2 = (H_1 \cup H_2) \sim (H_1 \cap H_2)$ is given by

$$C_{H_1 \triangle H_2} = (C_{H_1} + C_{H_2}) - (C_{H_1} \times C_{H_2})\surd$$

THEOREM **7.3**

Let $G = (X, \Gamma)$ be a finite graph, $\# X = n$. Then the characteristic matrix $\hat{C}$ of $\hat{\Gamma}$ is expressed by

$$\hat{C} = (I + C + C^2 + \cdots + C^{n-1})\surd$$

Proof. $\hat{\Gamma}x = \{x\} \cup \{\Gamma x\} \cup \{\Gamma^2 x\} \cup \cdots \cup \{\Gamma^{n-1}x\}$, since any simple path of G contains at most $n - 1$ arcs. Since I is the characteristic matrix of X, repeated application of Proposition 7.2 yields the result. ∎

COROLLARY **7.4**

$\hat{C} = (I + C)^{n-1}\surd$.

Proof. We use induction on $n = \#X$. The result is obvious for $n \leq 2$. If $n = 3$,

$$\begin{aligned}(I + C)^2\surd &= (I + C)(I + C)\surd = (I^2 + CI + IC + C^2)\surd \\ &= (I^2 + 2C + C^2)\surd = (I + C + C^2)\surd\end{aligned}$$

Assume the result for $n = k$. Then

$$\begin{aligned}(I + C)^k\surd &= (I + C)^{k-1}(I + C)\surd \\ &= (I + C + C^2 + \cdots + C^{k-1})(I + C)\surd \\ &= (I + C + C^2 + \cdots + C^{k+1}) + (C + C^2 + \cdots + C^k)\surd \\ &= (I + 2C + 2C^2 + \cdots + 2C^{k-1} + C^k)\surd \\ &= (I + C + C^2 + \cdots + C^k)\surd\end{aligned}$$

∎

THEOREM **7.5**

Let C be the connectivity matrix for a finite directed graph $G = (X, \Gamma)$. Consider the matrices $C^k = (c_{ij}^k)$, $k \geq 1$. Then c_{ii}^k, $i = 1, 2, \ldots, n$, is precisely equal to the number of circuits of length k involving the node x_i.

Proof. The result follows directly from Theorem 7.1 and the fact that x_i is in a circuit of length k if and only if there exists a path of length k from x_i. We note that such circuits need not be prime cycles, nor need they be simple, since $c_{ii}^k \neq 0$ implies that $c_{ii}^{pk} \neq 0$ for $p > 0$ (since one may traverse a cycle of length k starting at x_i p times ending up at x_i). ∎

7.3. THE DISTANCE MATRIX

DEFINITION **7.6**

We define a *geodesic from* x_i *to* x_j as a path of minimal length connecting x_i to x_j, and we denote by $\delta(x_i, x_j)$ [or $\delta(i, j)$ if the context is clear] the length of that path. We deliberately define $\delta(i, i) > 0$ in this context.

The geodesic measure δ is not a metric on a directed graph, for not only is $\delta(i, i) \neq 0$ but, in general $\delta(i, j) \neq \delta(j, i)$. However, we have that

$$\delta(i, j) + \delta(j, k) \geq \delta(i, k),$$

provided we define $\delta(i, j) = \infty$ if $x_j \notin \hat{\Gamma} x_i$.

A matrix $D = (d_{ij})$ called the *geodesic matrix* (or *distance matrix*) in which $d_{ij} = \delta(i, j)$. D may be constructed by the following algorithm.

ALGORITHM **7.7**

For a finite graph $G = (X, \Gamma)$ with connectivity matrix $C = (c_{ij})$, the geodesic matrix $D = (d_{ij})$ is constructed by setting

$$d_{ij} = \begin{cases} \min k \text{ such that } c_{ij}^k \neq 0 \\ \infty \qquad \text{if } c_{ij}^k = 0 \text{ for all } k \end{cases}$$

Proof. Follows immediately from Theorem 7.1. ∎

Example 7.8

Consider the graph in Fig. 7.1, whose connectivity matrix C is

$$C = \begin{bmatrix} 0 & 1 & 1 & 0 & 0 & 0 & 0 & 0 & 0 & 0 & 0 \\ 0 & 0 & 0 & 1 & 1 & 0 & 0 & 0 & 0 & 0 & 0 \\ 0 & 0 & 0 & 0 & 0 & 1 & 1 & 0 & 0 & 0 & 0 \\ 0 & 1 & 0 & 0 & 1 & 0 & 0 & 1 & 0 & 0 & 0 \\ 0 & 0 & 1 & 0 & 0 & 0 & 0 & 1 & 0 & 0 & 0 \\ 0 & 0 & 0 & 0 & 0 & 0 & 1 & 0 & 1 & 0 & 0 \\ 0 & 0 & 1 & 0 & 0 & 0 & 0 & 0 & 1 & 0 & 0 \\ 1 & 0 & 1 & 0 & 0 & 0 & 0 & 0 & 0 & 0 & 0 \\ 0 & 0 & 0 & 0 & 0 & 0 & 0 & 0 & 0 & 0 & 0 \\ 1 & 1 & 0 & 0 & 0 & 0 & 0 & 1 & 0 & 1 \\ 0 & 0 & 0 & 0 & 0 & 0 & 0 & 0 & 0 & 0 & 0 \end{bmatrix}$$

We first compute $\hat{C} = (I + C)^{n-1}\surd$:

$$(I + C)\surd = \begin{bmatrix} 1 & 1 & 1 & 0 & 0 & 0 & 0 & 0 & 0 & 0 & 0 \\ 0 & 1 & 0 & 1 & 1 & 0 & 0 & 0 & 0 & 0 & 0 \\ 0 & 0 & 1 & 0 & 0 & 1 & 1 & 0 & 0 & 0 & 0 \\ 0 & 1 & 0 & 1 & 1 & 0 & 0 & 1 & 0 & 0 & 0 \\ 0 & 0 & 1 & 0 & 1 & 0 & 0 & 1 & 0 & 0 & 0 \\ 0 & 0 & 0 & 0 & 0 & 1 & 1 & 0 & 1 & 0 & 0 \\ 0 & 0 & 1 & 0 & 0 & 0 & 1 & 0 & 1 & 0 & 0 \\ 1 & 0 & 1 & 0 & 0 & 0 & 0 & 1 & 0 & 0 & 0 \\ 0 & 0 & 0 & 0 & 0 & 0 & 0 & 0 & 1 & 0 & 0 \\ 1 & 1 & 0 & 0 & 0 & 0 & 0 & 0 & 1 & 1 & 1 \\ 0 & 0 & 0 & 0 & 0 & 0 & 0 & 0 & 0 & 0 & 1 \end{bmatrix}$$

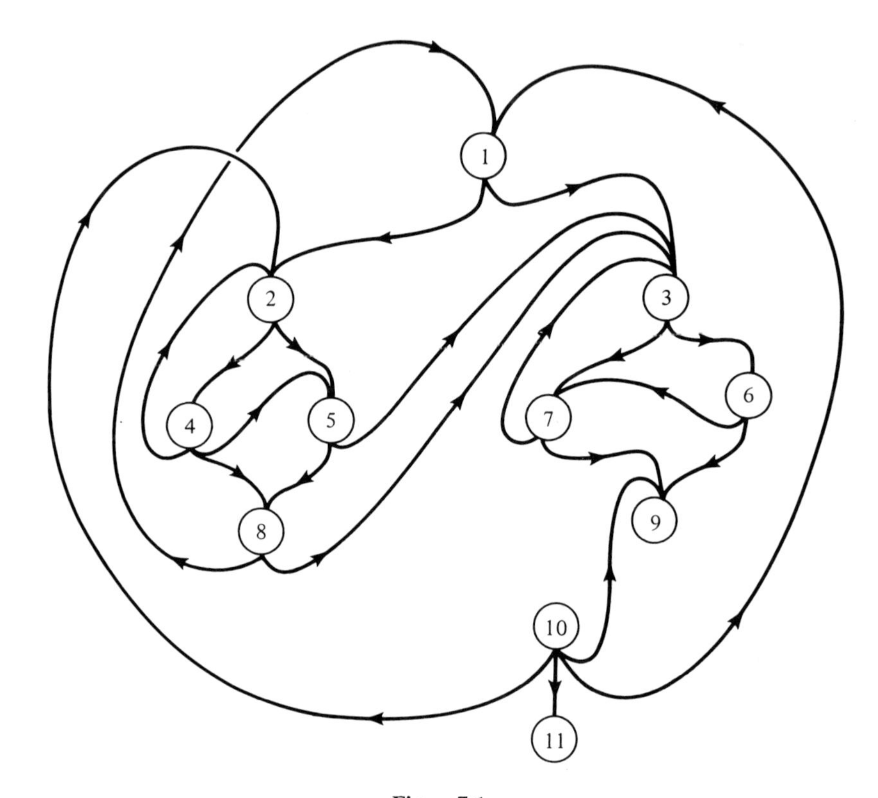

Figure 7.1

Computation yields

$$\hat{C} = (I + C)^{10}\surd = \begin{bmatrix} 1 & 1 & 1 & 1 & 1 & 1 & 1 & 1 & 1 & 0 & 0 \\ 1 & 1 & 1 & 1 & 1 & 1 & 1 & 1 & 1 & 0 & 0 \\ 0 & 0 & 1 & 0 & 0 & 1 & 1 & 0 & 1 & 0 & 0 \\ 1 & 1 & 1 & 1 & 1 & 1 & 1 & 1 & 1 & 0 & 0 \\ 1 & 1 & 1 & 1 & 1 & 1 & 1 & 1 & 1 & 0 & 0 \\ 0 & 0 & 1 & 0 & 0 & 1 & 1 & 0 & 1 & 0 & 0 \\ 0 & 0 & 1 & 0 & 0 & 1 & 1 & 0 & 1 & 0 & 0 \\ 1 & 1 & 1 & 1 & 1 & 1 & 1 & 1 & 1 & 0 & 0 \\ 0 & 0 & 0 & 0 & 0 & 0 & 0 & 0 & 1 & 0 & 0 \\ 1 & 1 & 1 & 1 & 1 & 1 & 1 & 1 & 1 & 1 & 1 \\ 0 & 0 & 0 & 0 & 0 & 0 & 0 & 0 & 0 & 0 & 1 \end{bmatrix}$$

The distance matrix may be computed by exponentiating C. (The results shown were obtained by computer rather than by hand, and we substituted 121 [$= 11^2 = (\#X)^2$] for ∞. Otherwise Algorithm 7.7 was faithfully followed.) From the matrix C we have a first matrix D_1:

$$D_1 = \begin{bmatrix} 121 & 1 & 1 & 121 & 121 & 121 & 121 & 121 & 121 & 121 & 121 \\ 121 & 121 & 121 & 1 & 1 & 121 & 121 & 121 & 121 & 121 & 121 \\ 121 & 121 & 121 & 121 & 121 & 1 & 1 & 121 & 121 & 121 & 121 \\ 121 & 1 & 121 & 121 & 1 & 121 & 121 & 1 & 121 & 121 & 121 \\ 121 & 121 & 1 & 121 & 121 & 121 & 121 & 1 & 121 & 121 & 121 \\ 121 & 121 & 121 & 121 & 121 & 121 & 1 & 121 & 1 & 121 & 121 \\ 121 & 121 & 1 & 121 & 121 & 121 & 121 & 121 & 1 & 121 & 121 \\ 1 & 121 & 1 & 121 & 121 & 121 & 121 & 121 & 121 & 121 & 121 \\ 121 & 121 & 121 & 121 & 121 & 121 & 121 & 121 & 121 & 121 & 121 \\ 1 & 1 & 121 & 121 & 121 & 121 & 121 & 121 & 1 & 121 & 1 \\ 121 & 121 & 121 & 121 & 121 & 121 & 121 & 121 & 121 & 121 & 121 \end{bmatrix}$$

$$C^2\surd = \begin{bmatrix} 0 & 0 & 0 & 1 & 1 & 1 & 1 & 0 & 0 & 0 & 0 \\ 0 & 1 & 1 & 0 & 1 & 0 & 0 & 1 & 0 & 0 & 0 \\ 0 & 0 & 1 & 0 & 0 & 0 & 1 & 0 & 1 & 0 & 0 \\ 1 & 0 & 1 & 1 & 1 & 0 & 0 & 1 & 0 & 0 & 0 \\ 1 & 0 & 1 & 0 & 0 & 1 & 1 & 0 & 0 & 0 & 0 \\ 0 & 0 & 1 & 0 & 0 & 0 & 0 & 0 & 1 & 0 & 0 \\ 0 & 0 & 0 & 0 & 0 & 1 & 1 & 0 & 0 & 0 & 0 \\ 0 & 1 & 1 & 0 & 0 & 1 & 1 & 0 & 0 & 0 & 0 \\ 0 & 0 & 0 & 0 & 0 & 0 & 0 & 0 & 0 & 0 & 0 \\ 0 & 1 & 1 & 1 & 1 & 0 & 0 & 0 & 0 & 0 & 0 \\ 0 & 0 & 0 & 0 & 0 & 0 & 0 & 0 & 0 & 0 & 0 \end{bmatrix}.$$

$$D_2 = \begin{bmatrix} 121 & 1 & 1 & 2 & 2 & 2 & 2 & 121 & 121 & 121 & 121 \\ 121 & 2 & 2 & 1 & 1 & 121 & 121 & 2 & 121 & 121 & 121 \\ 121 & 121 & 2 & 121 & 121 & 1 & 1 & 121 & 2 & 121 & 121 \\ 2 & 1 & 2 & 2 & 1 & 121 & 121 & 1 & 121 & 121 & 121 \\ 2 & 121 & 1 & 121 & 121 & 2 & 2 & 1 & 121 & 121 & 121 \\ 121 & 121 & 2 & 121 & 121 & 121 & 1 & 121 & 1 & 121 & 121 \\ 121 & 121 & 1 & 121 & 121 & 2 & 2 & 121 & 1 & 121 & 121 \\ 1 & 2 & 1 & 121 & 121 & 2 & 2 & 121 & 121 & 121 & 121 \\ 121 & 121 & 121 & 121 & 121 & 121 & 121 & 121 & 121 & 121 & 121 \\ 1 & 1 & 2 & 2 & 2 & 121 & 121 & 121 & 1 & 121 & 1 \\ 121 & 121 & 121 & 121 & 121 & 121 & 121 & 121 & 121 & 121 & 121 \end{bmatrix}$$

$$(C^3)\surd = \begin{bmatrix} 0&1&1&0&1&0&1&1&1&0&0\\ 1&0&1&1&1&1&1&1&0&0&0\\ 0&0&1&0&0&1&1&0&1&0&0\\ 1&1&1&0&1&1&1&1&0&0&0\\ 0&1&1&0&0&1&1&0&1&0&0\\ 0&0&0&0&0&1&1&0&0&0&0\\ 0&0&1&0&0&0&1&0&1&0&0\\ 0&0&1&1&1&1&1&0&1&0&0\\ 0&0&0&0&0&0&0&0&0&0&0\\ 0&1&1&1&1&1&1&1&0&0&0\\ 0&0&0&0&0&0&0&0&0&0&0 \end{bmatrix}$$

$$D_3 = \begin{bmatrix} 121&1&1&2&2&2&2&3&3&121&121\\ 3&2&2&1&1&3&3&2&121&121&121\\ 121&121&2&121&121&1&1&121&2&121&121\\ 2&1&2&2&1&3&3&1&121&121&121\\ 2&3&1&121&121&2&2&1&3&121&121\\ 121&121&2&121&121&3&1&121&1&121&121\\ 121&121&1&121&121&2&2&121&1&121&121\\ 1&2&1&3&3&2&2&121&3&121&121\\ 121&121&121&121&121&121&121&121&121&121&121\\ 1&1&2&2&2&3&3&3&1&121&1\\ 121&121&121&121&121&121&121&121&121&121&121 \end{bmatrix}$$

$$(C^4)\surd = \begin{bmatrix} 1&0&1&1&1&1&1&1&1&0&0\\ 1&1&1&0&1&1&1&1&1&0&0\\ 0&0&1&0&0&1&1&0&1&0&0\\ 1&1&1&1&1&1&1&1&1&0&0\\ 0&0&1&1&1&1&1&0&1&0&0\\ 0&0&1&0&0&0&1&0&1&0&0\\ 0&0&1&0&0&1&1&0&1&0&0\\ 0&1&1&0&1&1&1&1&1&0&0\\ 0&0&0&0&0&0&0&0&0&0&0\\ 1&1&1&1&1&1&1&1&1&0&0\\ 0&0&0&0&0&0&0&0&0&0&0 \end{bmatrix}$$

$$D_4 = \begin{bmatrix} 4 & 1 & 1 & 2 & 2 & 2 & 2 & 3 & 3 & 121 & 121 \\ 3 & 2 & 2 & 1 & 1 & 3 & 3 & 2 & 4 & 121 & 121 \\ 121 & 121 & 2 & 121 & 121 & 1 & 1 & 121 & 2 & 121 & 121 \\ 2 & 1 & 2 & 2 & 1 & 3 & 3 & 1 & 4 & 121 & 121 \\ 2 & 3 & 1 & 4 & 4 & 2 & 2 & 1 & 3 & 121 & 121 \\ 121 & 121 & 2 & 121 & 121 & 3 & 1 & 121 & 1 & 121 & 121 \\ 121 & 121 & 1 & 121 & 121 & 2 & 2 & 121 & 1 & 121 & 121 \\ 1 & 2 & 1 & 3 & 3 & 2 & 2 & 4 & 3 & 121 & 121 \\ 121 & 121 & 121 & 121 & 121 & 121 & 121 & 121 & 121 & 121 & 121 \\ 1 & 1 & 2 & 2 & 2 & 3 & 3 & 3 & 1 & 121 & 1 \\ 121 & 121 & 121 & 121 & 121 & 121 & 121 & 121 & 121 & 121 & 121 \end{bmatrix}$$

We find that we may stop at this point, since the 1s of $\hat{C}$ are in bijective correspondence with the finite entries (i.e., those less than 121) of D_4. Hence

$$D = \begin{bmatrix} 4 & 1 & 1 & 2 & 2 & 2 & 2 & 3 & 3 & 121 & 121 \\ 3 & 2 & 2 & 1 & 1 & 3 & 3 & 2 & 4 & 121 & 121 \\ 121 & 121 & 2 & 121 & 121 & 1 & 1 & 121 & 2 & 121 & 121 \\ 2 & 1 & 2 & 2 & 1 & 3 & 3 & 1 & 4 & 121 & 121 \\ 2 & 3 & 1 & 4 & 4 & 2 & 2 & 1 & 3 & 121 & 121 \\ 121 & 121 & 2 & 121 & 121 & 3 & 1 & 121 & 1 & 121 & 121 \\ 121 & 121 & 1 & 121 & 121 & 2 & 2 & 121 & 1 & 121 & 121 \\ 1 & 2 & 1 & 3 & 3 & 2 & 2 & 4 & 3 & 121 & 121 \\ 121 & 121 & 121 & 121 & 121 & 121 & 121 & 121 & 121 & 121 & 121 \\ 1 & 1 & 2 & 2 & 2 & 3 & 3 & 3 & 1 & 121 & 1 \\ 121 & 121 & 121 & 121 & 121 & 121 & 121 & 121 & 121 & 121 & 121 \end{bmatrix}$$

It is of interest to note that $(C^5)\surd$ is distinct from $(C^i)\surd$, $i = 1, 2, 3, 4$. Indeed,

$$(C^5)\surd = \begin{bmatrix} 1 & 1 & 1 & 0 & 1 & 1 & 1 & 1 & 1 & 0 & 0 \\ 1 & 1 & 1 & 1 & 1 & 1 & 1 & 1 & 1 & 0 & 0 \\ 0 & 0 & 1 & 0 & 0 & 1 & 1 & 0 & 1 & 0 & 0 \\ 1 & 1 & 1 & 1 & 1 & 1 & 1 & 1 & 1 & 0 & 0 \\ 0 & 1 & 1 & 0 & 1 & 1 & 1 & 1 & 1 & 0 & 0 \\ 0 & 0 & 1 & 0 & 0 & 1 & 1 & 0 & 1 & 0 & 0 \\ 0 & 0 & 1 & 0 & 0 & 1 & 1 & 0 & 1 & 0 & 0 \\ 1 & 0 & 1 & 1 & 1 & 1 & 1 & 1 & 1 & 0 & 0 \\ 0 & 0 & 0 & 0 & 0 & 0 & 0 & 0 & 0 & 0 & 0 \\ 1 & 1 & 1 & 1 & 1 & 1 & 1 & 1 & 1 & 0 & 0 \\ 0 & 0 & 0 & 0 & 0 & 0 & 0 & 0 & 0 & 0 & 0 \end{bmatrix}$$

If the vertex x_i is in a circuit of length l and $x_j \in \Gamma x_i$, it follows that if $(c_{ij}^k)\surd = 1$, then also $(c_{ij}^{k+ml})\surd = 1$, $m = 0, 1, 2, \cdots$. Moreover, if there are simple paths of length l and l' in G, with $l + 1 < l' \leq n - 1$, and no paths of length l'', $l < l'' < l'$, then $D_l = D_{l+1} = \cdots = D_{l'-1}$, while $D_{l'-1} \neq D_{l'}$. Hence it is not possible to determine the termination of Algorithm 7.8 until either $(C^{n-1})\surd$ or $\hat{C}$ has been computed or all of the entries of D_k are finite for some $k < n$.

We note from either $\hat{C}$ or D that, e.g., there are no paths from x_6 to x_4, there are no paths to either x_{10} or x_{11} from any other node, and the shortest path from x_5 to x_4 is of length 4.

7.4. FINDING SHORTEST PATHS

Some reasonable questions that may be asked at this point include: What is the shortest path from x_2 to x_9? How many paths of length 4 exist connecting x_2 and x_9?

Theorem 7.1 may be used to answer the latter question, for c_{29}^4 is equal to the number of paths of length 4 connecting x_2 and x_9. We start by listing the first 11 powers of C (under standard matrix algebra) for reference:

$$C^2 = \begin{bmatrix} 0 & 0 & 0 & 1 & 1 & 1 & 1 & 0 & 0 & 0 & 0 \\ 0 & 1 & 1 & 0 & 1 & 0 & 0 & 2 & 0 & 0 & 0 \\ 0 & 0 & 1 & 0 & 0 & 0 & 1 & 0 & 2 & 0 & 0 \\ 1 & 0 & 2 & 1 & 1 & 0 & 0 & 1 & 0 & 0 & 0 \\ 1 & 0 & 1 & 0 & 0 & 1 & 1 & 0 & 0 & 0 & 0 \\ 0 & 0 & 1 & 0 & 0 & 0 & 0 & 0 & 1 & 0 & 0 \\ 0 & 0 & 0 & 0 & 0 & 1 & 1 & 0 & 0 & 0 & 0 \\ 0 & 1 & 1 & 0 & 0 & 1 & 1 & 0 & 0 & 0 & 0 \\ 0 & 0 & 0 & 0 & 0 & 0 & 0 & 0 & 0 & 0 & 0 \\ 0 & 1 & 1 & 1 & 1 & 0 & 0 & 0 & 0 & 0 & 0 \\ 0 & 0 & 0 & 0 & 0 & 0 & 0 & 0 & 0 & 0 & 0 \end{bmatrix}$$

$$C^3 = \begin{bmatrix} 0 & 1 & 2 & 0 & 1 & 0 & 1 & 2 & 2 & 0 & 0 \\ 2 & 0 & 3 & 1 & 1 & 1 & 1 & 1 & 0 & 0 & 0 \\ 0 & 0 & 1 & 0 & 0 & 1 & 1 & 0 & 1 & 0 & 0 \\ 1 & 2 & 3 & 0 & 1 & 2 & 2 & 2 & 0 & 0 & 0 \\ 0 & 1 & 2 & 0 & 0 & 1 & 2 & 0 & 2 & 0 & 0 \\ 0 & 0 & 0 & 0 & 0 & 1 & 1 & 0 & 0 & 0 & 0 \\ 0 & 0 & 1 & 0 & 0 & 0 & 1 & 0 & 2 & 0 & 0 \\ 0 & 0 & 1 & 1 & 1 & 1 & 2 & 0 & 2 & 0 & 0 \\ 0 & 0 & 0 & 0 & 0 & 0 & 0 & 0 & 0 & 0 & 0 \\ 0 & 1 & 1 & 1 & 2 & 1 & 1 & 2 & 0 & 0 & 0 \\ 0 & 0 & 0 & 0 & 0 & 0 & 0 & 0 & 0 & 0 & 0 \end{bmatrix}$$

$$C^4 = \begin{bmatrix} 2 & 0 & 4 & 1 & 1 & 2 & 2 & 1 & 1 & 0 & 0 \\ 1 & 3 & 5 & 0 & 1 & 3 & 4 & 2 & 2 & 0 & 0 \\ 0 & 0 & 1 & 0 & 0 & 1 & 2 & 0 & 2 & 0 & 0 \\ 2 & 1 & 6 & 2 & 2 & 3 & 5 & 1 & 4 & 0 & 0 \\ 0 & 0 & 2 & 1 & 1 & 2 & 3 & 0 & 3 & 0 & 0 \\ 0 & 0 & 1 & 0 & 0 & 0 & 1 & 0 & 2 & 0 & 0 \\ 0 & 0 & 1 & 0 & 0 & 1 & 1 & 0 & 1 & 0 & 0 \\ 0 & 1 & 3 & 0 & 1 & 1 & 2 & 2 & 3 & 0 & 0 \\ 0 & 0 & 0 & 0 & 0 & 0 & 0 & 0 & 0 & 0 & 0 \\ 2 & 1 & 5 & 1 & 2 & 1 & 2 & 3 & 2 & 0 & 0 \\ 0 & 0 & 0 & 0 & 0 & 0 & 0 & 0 & 0 & 0 & 0 \end{bmatrix}$$

$$C^5 = \begin{bmatrix} 1 & 3 & 6 & 0 & 1 & 4 & 6 & 2 & 4 & 0 & 0 \\ 2 & 1 & 8 & 3 & 3 & 5 & 8 & 1 & 7 & 0 & 0 \\ 0 & 0 & 2 & 0 & 0 & 1 & 2 & 0 & 3 & 0 & 0 \\ 1 & 4 & 10 & 1 & 3 & 6 & 9 & 4 & 8 & 0 & 0 \\ 0 & 1 & 4 & 0 & 1 & 2 & 4 & 2 & 5 & 0 & 0 \\ 0 & 0 & 1 & 0 & 0 & 1 & 1 & 0 & 1 & 0 & 0 \\ 0 & 0 & 1 & 0 & 0 & 1 & 2 & 0 & 2 & 0 & 0 \\ 2 & 0 & 5 & 1 & 1 & 3 & 4 & 1 & 3 & 0 & 0 \\ 0 & 0 & 0 & 0 & 0 & 0 & 0 & 0 & 0 & 0 & 0 \\ 3 & 3 & 9 & 1 & 2 & 5 & 6 & 3 & 3 & 0 & 0 \\ 0 & 0 & 0 & 0 & 0 & 0 & 0 & 0 & 0 & 0 & 0 \end{bmatrix}$$

$$C^6 = \begin{bmatrix} 2 & 1 & 10 & 3 & 3 & 6 & 10 & 1 & 10 & 0 & 0 \\ 1 & 5 & 14 & 1 & 4 & 8 & 13 & 6 & 13 & 0 & 0 \\ 0 & 0 & 2 & 0 & 0 & 2 & 3 & 0 & 3 & 0 & 0 \\ 4 & 2 & 17 & 4 & 5 & 10 & 16 & 4 & 15 & 0 & 0 \\ 2 & 0 & 7 & 1 & 1 & 4 & 6 & 1 & 6 & 0 & 0 \\ 0 & 0 & 1 & 0 & 0 & 1 & 2 & 0 & 2 & 0 & 0 \\ 0 & 0 & 2 & 0 & 0 & 1 & 2 & 0 & 3 & 0 & 0 \\ 1 & 3 & 8 & 0 & 1 & 5 & 8 & 2 & 7 & 0 & 0 \\ 0 & 0 & 0 & 0 & 0 & 0 & 0 & 0 & 0 & 0 & 0 \\ 3 & 4 & 14 & 3 & 4 & 9 & 14 & 3 & 11 & 0 & 0 \\ 0 & 0 & 0 & 0 & 0 & 0 & 0 & 0 & 0 & 0 & 0 \end{bmatrix}$$

$$C^7 = \begin{bmatrix} 1 & 5 & 16 & 1 & 4 & 10 & 16 & 6 & 16 & 0 & 0 \\ 6 & 2 & 24 & 5 & 6 & 14 & 22 & 5 & 21 & 0 & 0 \\ 0 & 0 & 3 & 0 & 0 & 2 & 4 & 0 & 5 & 0 & 0 \\ 4 & 8 & 29 & 2 & 6 & 17 & 27 & 9 & 26 & 0 & 0 \\ 1 & 3 & 10 & 0 & 1 & 7 & 11 & 2 & 10 & 0 & 0 \\ 0 & 0 & 2 & 0 & 0 & 1 & 2 & 0 & 3 & 0 & 0 \\ 0 & 0 & 2 & 0 & 0 & 2 & 3 & 0 & 3 & 0 & 0 \\ 2 & 1 & 12 & 3 & 3 & 8 & 13 & 1 & 13 & 0 & 0 \\ 0 & 0 & 0 & 0 & 0 & 0 & 0 & 0 & 0 & 0 & 0 \\ 3 & 6 & 24 & 4 & 7 & 14 & 23 & 7 & 23 & 0 & 0 \\ 0 & 0 & 0 & 0 & 0 & 0 & 0 & 0 & 0 & 0 & 0 \end{bmatrix}$$

$$C^8 = \begin{bmatrix} 6 & 2 & 27 & 5 & 6 & 16 & 26 & 5 & 26 & 0 & 0 \\ 5 & 11 & 39 & 2 & 7 & 24 & 38 & 11 & 36 & 0 & 0 \\ 0 & 0 & 4 & 0 & 0 & 3 & 5 & 0 & 6 & 0 & 0 \\ 9 & 6 & 46 & 8 & 10 & 29 & 46 & 8 & 44 & 0 & 0 \\ 2 & 1 & 15 & 3 & 3 & 10 & 17 & 1 & 18 & 0 & 0 \\ 0 & 0 & 2 & 0 & 0 & 2 & 3 & 0 & 3 & 0 & 0 \\ 0 & 0 & 3 & 0 & 0 & 2 & 4 & 0 & 5 & 0 & 0 \\ 1 & 5 & 19 & 1 & 4 & 12 & 20 & 6 & 21 & 0 & 0 \\ 0 & 0 & 0 & 0 & 0 & 0 & 0 & 0 & 0 & 0 & 0 \\ 7 & 7 & 40 & 6 & 10 & 24 & 38 & 11 & 37 & 0 & 0 \\ 0 & 0 & 0 & 0 & 0 & 0 & 0 & 0 & 0 & 0 & 0 \end{bmatrix}$$

$$C^9 = \begin{bmatrix} 5 & 11 & 43 & 2 & 7 & 27 & 43 & 11 & 42 & 0 & 0 \\ 11 & 7 & 61 & 11 & 13 & 39 & 63 & 9 & 62 & 0 & 0 \\ 0 & 0 & 5 & 0 & 0 & 4 & 7 & 0 & 8 & 0 & 0 \\ 8 & 17 & 73 & 6 & 14 & 46 & 75 & 18 & 75 & 0 & 0 \\ 1 & 5 & 23 & 1 & 4 & 15 & 25 & 6 & 27 & 0 & 0 \\ 0 & 0 & 3 & 0 & 0 & 2 & 4 & 0 & 5 & 0 & 0 \\ 0 & 0 & 4 & 0 & 0 & 3 & 5 & 0 & 6 & 0 & 0 \\ 6 & 2 & 31 & 5 & 6 & 19 & 31 & 5 & 32 & 0 & 0 \\ 0 & 0 & 0 & 0 & 0 & 0 & 0 & 0 & 0 & 0 & 0 \\ 11 & 13 & 66 & 7 & 13 & 40 & 64 & 16 & 62 & 0 & 0 \\ 0 & 0 & 0 & 0 & 0 & 0 & 0 & 0 & 0 & 0 & 0 \end{bmatrix}$$

$$C^{10} = \begin{bmatrix} 11 & 7 & 66 & 11 & 13 & 43 & 70 & 9 & 70 & 0 & 0 \\ 9 & 22 & 96 & 7 & 18 & 61 & 100 & 24 & 102 & 0 & 0 \\ 0 & 0 & 7 & 0 & 0 & 5 & 9 & 0 & 11 & 0 & 0 \\ 18 & 14 & 115 & 17 & 23 & 73 & 119 & 20 & 121 & 0 & 0 \\ 6 & 2 & 36 & 5 & 6 & 23 & 38 & 5 & 40 & 0 & 0 \\ 0 & 0 & 4 & 0 & 0 & 3 & 5 & 0 & 6 & 0 & 0 \\ 0 & 0 & 5 & 0 & 0 & 4 & 7 & 0 & 8 & 0 & 0 \\ 5 & 11 & 48 & 2 & 7 & 31 & 50 & 11 & 50 & 0 & 0 \\ 0 & 0 & 0 & 0 & 0 & 0 & 0 & 0 & 0 & 0 & 0 \\ 16 & 18 & 104 & 13 & 20 & 66 & 106 & 20 & 104 & 0 & 0 \\ 0 & 0 & 0 & 0 & 0 & 0 & 0 & 0 & 0 & 0 & 0 \end{bmatrix}$$

$$C^{11} = \begin{bmatrix} 9 & 22 & 103 & 7 & 18 & 66 & 109 & 24 & 113 & 0 & 0 \\ 24 & 16 & 151 & 22 & 29 & 96 & 157 & 25 & 161 & 0 & 0 \\ 0 & 0 & 9 & 0 & 0 & 7 & 12 & 0 & 14 & 0 & 0 \\ 20 & 35 & 180 & 14 & 31 & 115 & 188 & 40 & 192 & 0 & 0 \\ 5 & 11 & 55 & 2 & 7 & 36 & 59 & 11 & 61 & 0 & 0 \\ 0 & 0 & 5 & 0 & 0 & 4 & 7 & 0 & 8 & 0 & 0 \\ 0 & 0 & 7 & 0 & 0 & 5 & 9 & 0 & 11 & 0 & 0 \\ 11 & 7 & 73 & 11 & 13 & 48 & 79 & 9 & 81 & 0 & 0 \\ 0 & 0 & 0 & 0 & 0 & 0 & 0 & 0 & 0 & 0 & 0 \\ 20 & 29 & 162 & 18 & 31 & 104 & 170 & 33 & 172 & 0 & 0 \\ 0 & 0 & 0 & 0 & 0 & 0 & 0 & 0 & 0 & 0 & 0 \end{bmatrix}$$

We note that $c_{49}^{4} = 4$. To enumerate the paths of length 4 connecting x_4 to x_9, we follow a technique of C. Flament. Clearly, the set Q of all nodes on simple paths connecting x_4 to x_9 is given by

$$Q = \{x_4, x_9\} \cup \{x_j \in X \mid \delta(4, j) + \delta(j, 9) = \delta(4, 9) = 4\}$$

Q may be found by juxtaposition of D_4 and $D_{,9}$, i.e.,

$$\begin{array}{r} D_4 = 2 \quad 1 \quad 2 \quad 2 \quad 1 \quad 3 \quad 3 \quad 1 \quad 4 \quad \infty \quad \infty \\ D_{,9} = 3 \quad 4 \quad 2 \quad 4 \quad 3 \quad 1 \quad 1 \quad 3 \quad \infty \quad 1 \quad \infty \\ \hline D_4 + D_{,9} = 5 \quad 5 \quad \underline{4} \quad 6 \quad \underline{4} \quad \underline{4} \quad \underline{4} \quad \underline{4} \quad \infty \quad \infty \quad \infty \end{array}$$

Hence $Q = \{x_4, x_9, x_3, x_5, x_6, x_7, x_8\}$.

Next we arrange the nodes of Q in columns such that the first column is x_4, and the nodes in the $(k + 1)$st column are of distance k from x_4. Finally, the appropriate connecting arcs are drawn (see Fig. 7.2). The enumeration of the paths is now straightforward:

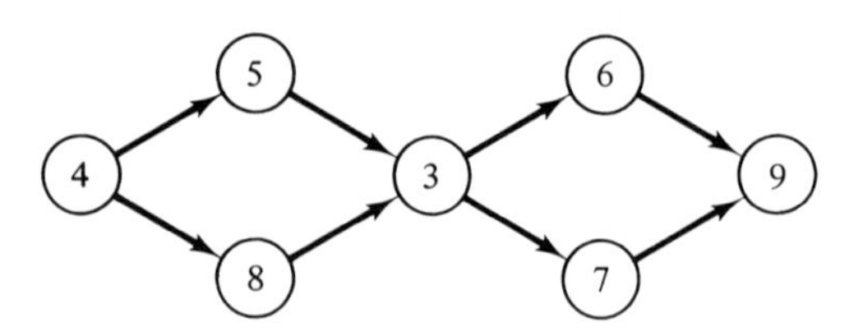

Figure 7.2

$$\mu_1 = [x_4, x_5, x_3, x_6, x_9]$$
$$\mu_2 = [x_4, x_5, x_3, x_7, x_9]$$
$$\mu_3 = [x_4, x_8, x_3, x_6, x_9]$$
$$\mu_4 = [x_4, x_8, x_3, x_7, x_9]$$

Flament's technique only produces simple paths. For example, we see that $c_{43}^5 = 10$, while $\delta(4, 3) = 2$. The 10 paths are

$$\mu_1 = [x_4, x_2, x_5, x_8, x_1, x_3]$$
$$\mu_2 = [x_4, x_2, x_5, x_3, x_7, x_3]$$
$$\mu_3 = [x_4, x_2, x_4, x_2, x_5, x_3]$$
$$\mu_4 = [x_4, x_2, x_4, x_5, x_8, x_3]$$
$$\mu_5 = [x_4, x_2, x_4, x_8, x_1, x_3]$$
$$\mu_6 = [x_4, x_5, x_3, x_6, x_7, x_3]$$
$$\mu_7 = [x_4, x_8, x_1, x_2, x_5, x_3]$$
$$\mu_8 = [x_4, x_8, x_1, x_3, x_7, x_3]$$
$$\mu_9 = [x_4, x_8, x_3, x_6, x_7, x_3]$$
$$\mu_{10} = [x_4, x_5, x_8, x_3, x_7, x_3]$$

Only μ_1, μ_2, and μ_7 are simple paths; each of the others involves some cycle. Since G is not weak, there are infinitely many paths in the graph and the sequence $C, C^2, C^3, \ldots$ does not have a limit.

PROPOSITION **7.9**

Let $G = (X, \Gamma)$ be cycle-free, and let C be the associated connectivity matrix. Then the sequence $C, C^2, C^3, \ldots$ has a limit. In fact,

$$\lim_k C^k = C^n = (0)$$

where $n = \#X$. There may exist an integer $p < n$ such that $C^p = (0)$.

Proof. Since G is cycle-free, no path may use more than $n - 1$ arcs. Hence there are no paths of length $\geq n$. By Theorem 7.1, $C^n = (0)$. Finally, there may be no paths of length $n - 1$, proving the second assertion. ∎

ALGORITHM **7.10** (*Flament*)

To find the geodesics connecting two vertices x_m, x_n in the finite directed graph $G = (X, \Gamma)$, with $D = (\delta_{ij})$ the distance matrix associated with G,

1. $\Delta_m = \Delta_{m0} = \{x_m\}$, $\Delta_n = \Delta_{m,\delta_{mn}}\{x_n\}$.

2. Define sets $\Delta_{mk} = \{x_p \in X | \delta_{mp} = k$ and $\delta_{mp} + \delta_{pn} = \delta_{mn}\}$, $k = 1, 2, \ldots, \delta_{mn} - 1$, $\Delta = \sum_{k=0}^{mn} \Delta_{mk}$.

3. Define $\Gamma_\Delta \Delta_{mk} = \Delta_{mk+1} \cap \Gamma\Delta_{mk}$, $k = 0, 1, 2, \ldots, \delta_{mn} - 1$.

4. Then every path in $G_\Delta = (\Delta, \Gamma_\Delta)$ connecting Δ_m to Δ_n is a geodesic in G connecting x_m to x_n.

Proof. It is clear that the Δ_{mk} are pairwise disjoint, so that the Δ_{mk} form a partition of G_Δ. Furthermore, there are no paths between two elements of any one Δ_{mk}, nor are there any arcs connecting nodes in Δ_{mk} to nodes in Δ_{mk} where $k' < k$. Hence any path of length δ_{mn} in Δ, starting in Δ_m, must touch an element of each Δ_{mk} and therefore end in $\Delta_n = \Delta_{m,\delta_{mn}}$. Moreover, the construction of the sets Δ_{mk} guarantees that the elements of each Δ_{mk} are on some geodesic from x_m to x_n. ∎

7.5. ENUMERATING PRIME CYCLES

Figure 6.5 possesses a prime cycle, $\mu = [b, c, d, e]$ which is not a geodesic, and so not every prime cycle of a graph can be found by employing Algorithm 7.10 to find the shortest paths from each x_i to itself. However, B. Roy's method of *elementary developments* can be used to enumerate the elementary cycles of G. From these, the prime cycles are easily detected.

Essentially, this method efficiently lists the elementary paths of the graph. We present it with an example finding the elementary cycles of Fig. 6.5.

ALGORITHM **7.12**

To enumerate the elementary paths ensuing from a node i in a graph $G = (X, \Gamma)$ with $\#X = n$,

1. Let $T(i)$ be an $(n - 1) \times (n - 1)$ tableau in which there is no row corresponding to i. For each $k \in \Gamma i$, let $T(i)_{k1} = i$.

2. Assuming that column $q - 1$ has been filled in, column q may be established by tentatively setting each $T(i)_{kq} = \{\Gamma^{-1}k\} \sim \{i\}$ and then suppressing

a. All h in column q such that $T(i)_{h,q-1}$ is empty.

b. All h in $T(i)_{kq}$ such that there does not exist a simple path $\mu[i, k]$ terminating in the arc (h, k).

The suppressions of 2.b are easily determined by enumerative backtracking from k to i.

Example 7.12

Consider the graph of Fig. 6.5. We have

$$\Gamma^{-1}a = \varnothing$$
$$\Gamma^{-1}b = \{a, e, g\}$$
$$\Gamma^{-1}c = \{b, f\}$$
$$\Gamma^{-1}d = \{c, g\}$$
$$\Gamma^{-1}e = \{a, d, h\}$$
$$\Gamma^{-1}f = \{b, c\}$$
$$\Gamma^{-1}g = \{f, e\}$$
$$\Gamma^{-1}h = \{e, g\}$$

We are interested in the prime cycles of $X \sim \{a\}$, and we start from, say, b:

$T(b)$

	1	2	3	4	5	6
c	b	f	f	f	f	f
d		c~~g~~	cg	cg	cg	cg
e		~~dh~~	dh	dh	dh	dh
f	b	c	c	c	c	c
g		~~e~~f	ef	ef	ef	ef
h		~~eg~~	eg	eg	eg	eg

We show $T(b)$ after suppressing elements from column 2. Column 3 immediately becomes

	3
c	f
d	cg
e	d~~h~~
f	c
g	~~e~~f
h	~~e~~g

since $T(b)_{e2}$ and $T(b)_{h2}$ are empty. We eliminate f from $T(b)_{c3}$, since $[b, c, f, c]$ is not elementary. Similarly, c is eliminated from $T(b)_{f3}$. We keep $T(b)_{g3}$, since $[b, c, f, g]$ is simple. We next obtain

	1	2	3	4	5	6
c	b	f		~~f~~	~~f~~	f
d		c	cg	~~c~~g	~~c~~g	cg
e			d	dh	dh	dh
f	b	c		~~c~~	~~c~~	c
g		f	f	e~~f~~	e~~f~~	ef
h			g	eg	eg	eg

since all remaining paths to nodes in column 4 are simple. In addition to the obvious suppressions in column 5, we eliminate g from $T(b)_{d5}$ since $[b, c, d, e, g, d]$ is not simple. We finally obtain

	1	2	3	4	5	6	
c	*b*	*f*					
d		*c*	*cg*	*g*			
e			*d*	*dh*	*dh*		*
f	*b*	*c*					
g		*f*	*f*	*e*	*e*		*
h			*g*	*eg*	*eg*	*eg*	

We place asterisks in the rows corresponding to immediate predecessors of b. The elementary cycles containing b are

$$\{b, f, g, b\}$$
$$\{b, c, d, e, b\}$$
$$\{b, c, f, g, b\}$$
$$\{b, f, c, d, e, b\}$$
$$\{b, f, g, d, e, b\}$$
$$\{b, f, g, h, e, b\}$$
$$\{b, c, d, e, g, b\}$$
$$\{b, c, f, g, d, e, b\}$$
$$\{b, c, f, g, h, e, b\}$$
$$\{b, f, c, d, e, g, b\}$$

We eliminate b from the graph and establish $T(c)$ directly from $T(b)$:

	1	2	3	4	5	
d	*c*	~~*g*~~	*g*			
e		*d*	~~*dh*~~	*dh*		
f	*c*					*
g		*f*	*e*	~~*e*~~		
h		~~*g*~~	*eg*	~~*e*~~*g*	~~*eg*~~	

Simple paths exist to each of d, c, f, g, and h, but the unique elementary cycle containing c is $[c, f, c]$. $T(d)$ is obtained by eliminating d from $T(c)$:

	1	2	3	
e	d		~~h~~	
f				
g		e		*
h		e~~g~~	g	

We obtain

$$\{a, e, g, d\}$$

$T(e)$

	1	2	
f			
g	e		
h	e	g	*

This produces

$$\{e, h, e\}$$

$$\{e, g, h, e\}$$

$T(f)$, $T(g)$, and $T(h)$ produce no new information, and so there are 14 elementary cycles in G.

THEOREM **7.13**

Let $G = (X, \Gamma)$ be a finite directed graph, with C its connectivity matrix, $\# X = n$. Let P be an $m \times n$ matrix with P_j the characteristic vector of one of the m intervals of C. Then $C_{\partial p}$, the connectivity matrix of the derived graph, is given by

$$C_{\partial p} = (PC \times \tilde{P})(P^T)\surd$$

Proof. The kth row of $PC\surd$ is the characteristic vector of ΓI_k, the kth interval of G, so that $(PC \times \tilde{P})_k\surd$ is the characteristic vector of $\Gamma I_k \sim I_k = \Gamma^{\S} I_k$. However, $\Gamma^{\S} I_k$ is a subset of the interval heads of G, since the only node in the interval with predecessors from outside the interval is the interval head. Hence $(PC \times \tilde{P})_k \times P_j^T$ has at most one 1, viz., the head of the interval I_j, so that the k, jth component of $C_{\partial p}$ is 1 if and only if there is an arc from I_k to the head of I_j. ▮

EXERCISE

Algorithm 7.11 can be modified as follows so that it produces the prime cycles of a graph.

To find the prime cycles containing node i, one juxtaposes D_i and $D_{,i}$, producing $D_{i} + D_{,i}$ as we did in Section 7.4. As in Algorithm 7.11, the prime cycles of length δ_{ii} may be enumerated.

Prove that if there exists a nonempty set $K_n = \{k \mid \delta_{ik} + \delta_{ki} = n > \delta_{ii}\}$ and if there exist paths $\mu = [i, k_{j_1}, k_{j_2}, \ldots, k_{j_m}, i]$ with $k_{j_p} \in K_n$ and $l(\mu) = n$, then μ is a prime cycle.

Prove that all of the prime cycles may be found in this manner.

PART II APPLICATIONS TO GLOBAL PROGRAM ANALYSIS

The principle is so general that no particular application of it is possible.

G. Polya
How To Solve It

"*You know my methods. Apply them.*"

Sir Arthur Conan Doyle
The Sign of the Four

OVERVIEW

In Part I we sketched those segments of the theory of directed graphs which are necessary to a global analysis of the structure of a computer program. In particular, we have placed emphasis on the detection of strong connectivity in finite directed graphs possessing a unique node with no predecessors.

A partition, the interval, was defined to permit a study of program segments having a lattice structure with a partial order called backdominance. It will be seen that the properties of the interval, in the original program graph and the associated derived graphs, are invoked throughout the global optimization process.

Various optimization strategies involve the introduction of new nodes to the original program graph, while others cause code to be "moved" from one part of the program to another. These transformations depend on our ability to guarantee that the transformed code executes at least as efficiently in approximately the same amount of computer store as the original program and that it obtains the same numerical results.

The general problem of proving two algorithms to be computationally equivalent has been shown elsewhere to be recursively unsolvable. It is, therefore, not our ambition to approach this problem from the viewpoint of automata theory but instead to be as pragmatic as is feasible. After all, we wish to produce a corpus of techniques that can be employed by a real compiler to optimize code. Clearly, much of the analysis undertaken by human programmers when they personally analyze their programs cannot only be automated but can be carried out on a far more extensive basis than is within the capabilities of a mortal.

No attempt is made in this treatment to suggest that compilers need employ artificial intelligence in order to "understand" the code or algorithms being optimized. Consequently, there need exist general criteria that can be employed to indicate the set of transformations which are to be applied to each block of code in the program in such a manner that computational equivalence is maintained. The implications of this requirement are that it be possible, at each step of the optimization process, to prove that the program will run to completion if and only if the optimized version of the program runs to completion with identical results.

It will be seen that there exist optimization transformations which, while desirable, cannot be employed due to a potential danger that they either cause the program to abort or, paradoxically, that they actually slow the program down due to certain dependencies of the program on its data. Indeed, because of these problems, we can only expect to produce an analysis of the program which includes "suggestions" to the programmer for code enhancements which can be made only with his insights to the program and its precision requirements.

As the title of this part indicates, the theory is more general than are its applications. Fortunately, the existence of successful optimizing compilers is an indication of the utility of the theory.

8 DATA FLOW ANALYSIS: DEPENDENCY AND REDUNDANCY EQUATIONS

8.1. ARITHMETIC EXPRESSIONS

The time is now ripe for us to scrutinize code and its suitability for optimization. We shall not be concerned with how indicated computations are performed on a computer except at a superficial level. Nor shall we strictly employ the conventions of any specific programming language or system. Instead, we shall investigate those trends that appear to be general and build a systematic approach to the global optimization process.

We shall start by studying computational (or arithmetic) expressions. These are quite analogous to the sums, products, and surds of secondary arithmetic. We shall denote by $+$, $-$, $\times$, $\div$, and $\uparrow$ the operations plus, minus, times, divide, and exponentiate. Parentheses may be employed to modify the operator precedence, which is otherwise assumed to be that of arithmetic. Normally, constants will be written as numbers, while variables will appear as lowercase Latin or Greek letters optionally subscripted or followed by digits or letters for mnemonic reasons.

8.1.1. Definitions of Variables

It is convenient to visualize variables as holding places in the computer's magnetic store. A variable is said to have been *defined* or *set* each time the contents of its holding place have been modified. For example, if x is a variable, x can be defined as, say, the sum of the variables u and v, written as

$$x \longleftarrow u + v$$

At some subsequent point in the program x can be given a new definition.

Say x is to be incremented by the constant 5. This would be written as

$$x \longleftarrow x + 5$$

and is interpreted to mean that the "holding place of x is defined to contain the sum of the contents of the holding place of x and the constant 5."

8.1.2. Auxiliary Variables

We note that two distinct events may be said to be taking place in the statement

$$x \longleftarrow u + v$$

The first is constructive; i.e., the sum of the contents of u and v is computed. The second is destructive in that the contents of the holding place of x are destroyed while being replaced by the new computation.

We observe the manner in which this phenomenon occurs to disadvantage in the code shown below:

$$\begin{array}{ll}
\vdots & \\
x \longleftarrow (u \times v) \uparrow w & (1) \\
y \longleftarrow x + 2 & (2) \\
x \longleftarrow y - y \uparrow 0.5 & (3) \\
w \longleftarrow (u \times v) \uparrow w & (4) \\
\vdots &
\end{array}$$

Here, had not x been redefined on line (3), w could simply have been defined as the contents of x, since the expressions on lines (1) and (4) are identical and u, v, and w remain unchanged. Unfortunately, the code on line (3) makes this impossible. However, if the code were rewritten as

$$\begin{array}{ll}
\vdots & \\
\alpha_1 \longleftarrow (u \times v) \uparrow w & (1') \\
x \longleftarrow \alpha_1 & (2') \\
\alpha_2 \longleftarrow \alpha_1 + 2 & (3') \\
y \longleftarrow \alpha_2 & (4') \\
\alpha_3 \longleftarrow \alpha_2 - \alpha_2 \uparrow 0.5 & (5')
\end{array}$$

$$x' \longleftarrow \alpha_3 \qquad (6')$$

$$w \longleftarrow \alpha_1 \qquad (7')$$

$$\vdots$$

we would observe that α_1 retained the value of x even though x was subsequently redefined. This permits us to reuse α_1 in the definition of w rather than recompute it. Clever scrutiny reveals that we could easily eliminate the definition of x on line (2′) since x is only to be redefined later. It might be desirable to move the definition of w from line (7′) to linc (2′) for reasons that will be discussed in Chapter 13 when we consider register allocation. The result of this optimization is the code

$$\vdots$$

$$w \longleftarrow (u \times v) \uparrow w \qquad (1'')$$

$$y \longleftarrow w + 2 \qquad (2'')$$

$$x \longleftarrow y - y \uparrow 0.5 \qquad (3'')$$

$$\vdots$$

The major gain from this example was the elimination of one computation, to wit, $(u \times v) \uparrow w$. We shall now undertake a study of the considerations necessary to make such improvements.

8.1.3. Criteria for Using Auxiliary Variables

We first note that the value $\alpha_1 \longleftarrow (u \times v) \uparrow w$ can be destroyed in three ways, viz., by a new definition of u, v, or w. If only w were redefined, $u \times v$ would remain unchanged, so that the entire quantity should not have to be recomputed. To attempt the conservation of previously computed quantities, we shall define a variable α_i, called a *compiler variable*, in which to hold the result of each *binary* computation. E.g., in the case of $(u \times v) \uparrow w$, we would have two definitions:

$$\alpha_i \longleftarrow u + v$$

$$\alpha_{i+1} \longleftarrow \alpha_i \uparrow w$$

Then, if w were redefined at some subsequent point while u and v retained their prior values, the new value of $(u \times v) \uparrow w$ would be simply

$$\alpha_{i+1} \longleftarrow \alpha_i \uparrow w$$

The variable α_i need not necessarily contain the same value throughout the program, even though it represents a unique binary expression. For, if α_i represents $u \times v$ and u or v is redefined, the quantity contained in α_i no longer equals the result of a fresh computation of $u \times v$. It would therefore be an error to attempt to use the value of α_i instead of recomputing $u \times v$ under such circumstances, and so some form of analysis will be required to ascertain when α_i is current and when it is not. This analysis is described in detail below.

The correspondence between the α_i and binary expressions need not be bijective, but may instead be many–one. For example, if α_i represents $u \times v$ and α_j represents $u \times w$, it *may* be desirable to have both α_i and α_j represent $u \times v$ when $v = w$, as shown in the following code:

$$
\begin{array}{c}
\vdots \\
x \longleftarrow u \times v \\
w \longleftarrow v \\
\vdots \\
r \longleftarrow u \times w \\
v \longleftarrow s \\
\vdots \\
t \longleftarrow u \times w \\
\vdots
\end{array}
$$

which could be idealized as

$$
\begin{array}{c}
\vdots \\
\alpha_1 \longleftarrow u \times v \\
x \longleftarrow \alpha_1 \\
w \longleftarrow v \\
\vdots \\
\alpha_2 \longleftarrow \alpha_1 \\
r \longleftarrow \alpha_1 \\
v \longleftarrow s
\end{array}
$$

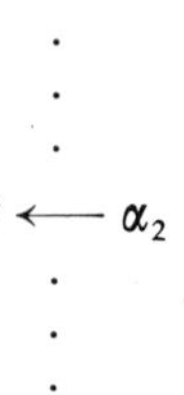

The definition of α_1 might later prove to have been unnecessary, so that α_2 could be substituted for α_1 from the first. In that case the generated code might resemble

$$\begin{array}{c} \vdots \\ \alpha_2 \leftarrow u \times v \\ x \leftarrow \alpha_2 \\ w \leftarrow v \\ \vdots \\ r \leftarrow \alpha_2 \\ v \leftarrow s \\ \vdots \\ t \leftarrow \alpha_2 \\ \vdots \end{array}$$

There might be other possible improvements to this code, but they are dependent on considerations we have not yet discussed. The important item in this example is that by idealistically identifying the same expression with two distinct variables, one copy of the expression remained valid after the other was destroyed.

8.2. CANONICAL REPRESENTATIONS OF ARITHMETIC EXPRESSIONS

To further enhance the likelihood of exposing formal identities, we shall adopt certain canonical representations, notably the following: if $\circ$ is a commutative, associative binary operator, the expression

$$w_1 \circ w_2 \circ \cdots \circ w_n$$

will be transformed into

$$(w_{\pi(1)} \circ w_{\pi(2)}) \circ (w_{\pi(3)} \circ w_{\pi(4)}) \circ \cdots \circ (w_{\pi(n-1)} \circ w_{\pi(n)}) \qquad (n \text{ even})$$

or

$$(w_{\pi(1)} \circ w_{\pi(2)}) \circ (w_{\pi(3)} \circ w_{\pi(4)}) \circ \cdots \circ (w_{\pi(n-2)} \circ w_{\pi(n-1)}) \circ w_{\pi(n)} \qquad (n \text{ odd})$$

where π is a permutation of $\{1, 2, \ldots, n\}$ such that the $w_{\pi(i)}$ are in lexicographic order.

This ordering is a compromise that reveals only a statistically small subset of the possible common subexpressions. For example, $w_2 \circ w_4$ is common to $w_1 \circ w_2 \circ w_4$ and $w_2 \circ w_3 \circ w_4$ but will not be exposed for elimination. However, because of the large number of possible pairings of the n variables in a program, it is clear that no scheme can expose all possible subexpressions in a practical manner.

The scheme proposed above is evidently superior to one in which terms are simply associated from the left, as is shown in the following contrived example. Assume the original code to be

$$\begin{aligned} w_1 &\longleftarrow w_2 \circ w_3 \circ w_4 \circ w_5 \circ w_6 \circ w_7 \\ w_3 &\longleftarrow f \circ g \\ h &\longleftarrow w_2 \circ w_3 \circ w_4 \circ w_5 \circ w_6 \circ w_7 \end{aligned}$$

Now, if the definition of w_1 were parsed as

$$\begin{aligned} \alpha_1 &\longleftarrow w_2 \circ w_3 \\ \alpha_2 &\longleftarrow \alpha_1 \circ w_4 \\ \alpha_3 &\longleftarrow \alpha_2 \circ w_5 \\ \alpha_4 &\longleftarrow \alpha_3 \circ w_6 \\ \alpha_5 &\longleftarrow \alpha_4 \circ w_7 \\ w_1 &\longleftarrow \alpha_5 \end{aligned}$$

the modification of w_3 would force recomputation of α_1, α_2, α_3, α_4, and α_5. If, on the other hand, the definition of w_1 were parsed as

$$\begin{aligned} \alpha_1 &\longleftarrow w_2 \circ w_3 \\ \alpha_2 &\longleftarrow w_4 \circ w_5 \\ \alpha_3 &\longleftarrow \alpha_1 \circ \alpha_2 \\ \alpha_4 &\longleftarrow w_6 \circ w_7 \\ \alpha_5 &\longleftarrow \alpha_3 \circ \alpha_4 \\ w_1 &\longleftarrow \alpha_5 \end{aligned}$$

only α_1, α_3, and α_5 would have to be computed for the definition of h.

Distributive operators pose a different problem. If $*$ is distributive over $\circ$ so that $\alpha * (\beta \circ \gamma) = (\alpha * \beta) \circ (\alpha * \gamma)$, the choice of a standard form is not easily determined. For if, say, $\alpha * \beta$ is available from a previous computation, $(\alpha * \beta) \circ (\alpha * \gamma)$ might be preferable to $\alpha * (\beta \circ \gamma)$.†

8.2.1. Machine-Dependant Restrictions

It should be noted that computer operations do not generally obey the associative and distributive laws of arithmetic. On some computers $(2 + 2^{-24}) + 2^{-24}$ is not equal to $2 + (2^{-24} + 2^{-24})$ due to a lack of precision in the representation of floating-point numbers. Similar examples are readily provided for the distributive law. Hence any schemes which take advantage of the algebraic properties of real numbers must be cognizant of the restrictions imposed by the computer's hardware.

There may also exist certain semantic restrictions imposed by the particular programming language being employed. For example, if $\vee$ and $\wedge$ are the logical operations *or* and *and*, respectively, it might be required in the expression $(\alpha \wedge \beta) \vee \gamma$ that $(\alpha \wedge \beta)$ is computed prior to the computation of γ, although *logically* $\vee$ is a commutative operator.

These problems will be examined more closely in Chapter 11. Meanwhile, it will be assumed that, by some means, expressions in the programming language are mapped into a computationally equivalent canonical pairing of operators and operands such that each such (binary) expression defines some (possibly compiler-generated) variable.

8.3. REDUNDANCY

Now let $\alpha \leftarrow a \circ b$ be an arbitrary binary expression occurring in the kth block of a program, and assume that $a \circ b$ appears at some subsequent point in the same block. We ask under what conditions the contents of α may be substituted for the latter computation of $a \circ b$? We shall say that a computation $a \circ b$ is *redundant* if its value is available in some variable α.

The answer to our question is simple: The latter computation of $a \circ b$ is identical to the contents of the variable α if and only if there does not occur a definition of either a or b between the definition of α and the subsequent appearance of $a \circ b$. We say that the definition of α is *killed* by a new

†Some interesting results on the typical expressions occuring in FORTRAN programs may be found in Knuth's study. (See Bibliography.)

definition of a or of b and refer to such a redefinition as a *definition relevant to* α.

We shall periodically make use of schematic diagrams of blocks within the program graph, and code within certain blocks will be indicated in proximity to those blocks. It will always be the case that only relevant code is shown, the remainder being suppressed.

Hence, by way of summary of the above characterization, $a \circ b$ is redundant in Fig. 8.1(a) and irredundant in Fig. 8.1(b).

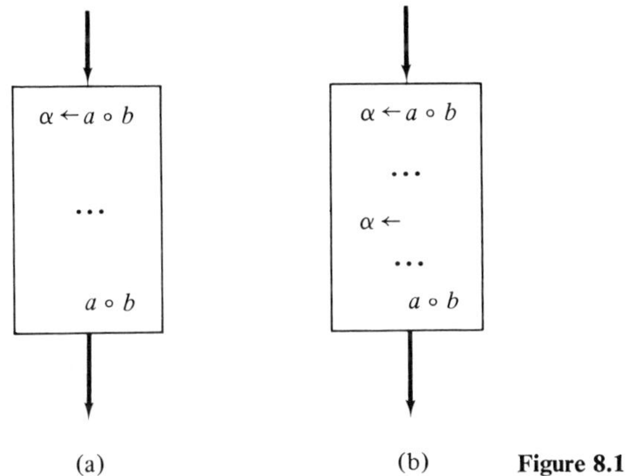

Figure 8.1

In Fig. 8.1, the idiom

$$a \longleftarrow$$

is used to indicate a redefinition of the variable a, while in both figures

$$a \circ b$$

standing alone indicates a use of the computation $a \circ b$.

8.3.1. Availability of Expressions

The question of the redundancy of subsequent occurrences of a given computation within a block has a straightforward answer. Next we shall proceed to generalize the solution to the program graph as a whole. To achieve this goal, we shall first define the concept of *availability on exit*.

DEFINITION **8.1**

The expression $\alpha \leftarrow a \circ b$ is *available at a point p* of a program provided that the value contained in the variable α is equal to the value of $a \circ b$ computed with the values of a and b at the point p.

DEFINITION **8.2**

The expression α is *available on exit* from block b if α is available at the last point within b.

DEFINITION **8.3**

The expression α is *available on entrance* to a block b if and only if α is available on exit from each immediate predecessor block of b.

Note that no expressions are available on entrance to the first block of the program, since that block, by definition, has no predecessors. These definitions immediately yield

PROPOSITION **8.4**

The expression α is available on exit from the block b if and only if either of the following conditions hold:

1. α is computed within block b and no relevant computations to α occur subsequent to the final computation of α in b.

2. α is available on entrance to block b and no relevant computations to α occur within block b.

8.3.2. Redundancy Equations

Proposition 8.4 can be expressed symbolically in terms of the Boolean lattice algebra of Chapter 6. Let

$$x_{\alpha b} = \begin{cases} 1 & \text{if } \alpha \text{ is available on entrance to } b \\ 0 & \text{otherwise} \end{cases}$$

$$k_{\alpha b} = \begin{cases} 1 & \text{if no relevant computation to } \alpha \text{ occurs in } b \\ 0 & \text{otherwise} \end{cases}$$

$$c_{\alpha b} = \begin{cases} 1 & \text{if } \alpha \text{ is computed in } b \text{ and no relevant computation to } \alpha \\ & \text{follows the last definition of } \alpha \text{ in } b \\ 0 & \text{otherwise} \end{cases}$$

Then the availability of α on exit from b, $y_{\alpha b}$ is given by

COROLLARY **8.5**

$$y_{\alpha b} = k_{\alpha b} x_{\alpha b} + c_{\alpha b}.$$

In general, a consistent system of Boolean equations has a multiplicity of

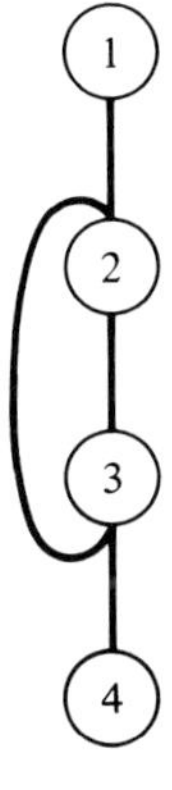

Figure 8.2

solutions. For example, the graph in Fig. 8.2 yields the equations (for an unspecified expression α)

$$y_1 = k_1 x_1 + c_1$$
$$y_2 = k_2 y_1 y_3 + c_2$$
$$y_3 = k_3 y_2 + c_3$$
$$x_4 = y_3$$

Since $x_1 = 0$ by hypothesis (no computations are available on entrance to the first block of the program), substitution yields

$$y_2 = c_1 k_2 k_3 y_2 + k_2 c_1 c_3 + c_2$$

If $c_2 = c_3 = 0$ but $k_2 = k_3 = 1$, we obtain

$$y_2 = c_1 y_2$$
$$y_3 = y_2$$
$$x_4 = y_2$$

independent of the value of c_1. Indeed, if $c_1 = 1$, we may obtain $x_4 = 1$ or $x_4 = 0$.

The physical interpretation of this situation is that the expression α is neither killed nor computed in the cycle [2, 3] but is available on exit from block 1 ($c_1 = 1$). Then we should have that α is available on entrance to block 4 (the solution set $y_1 = y_2 = y_3 = x_4 = 1$), but we could obtain the solution $y_1 = 1$, $y_2 = y_3 = y_4 = 0$, which indicates the necessity of computing α unnecessarily in block 4.

8.4. SOLUTIONS TO THE REDUNDANCY EQUATIONS

It is clear that all solutions $(y_{\alpha 1}, y_{\alpha 2}, \ldots, y_{\alpha n})$ must lie between the bounds $(k_{\alpha 1} + c_{\alpha 1}, k_{\alpha 2} + c_{\alpha 2}, \ldots, k_{\alpha n} + c_{\alpha n})$ and $(c_{\alpha 1}, c_{\alpha 2}, \ldots, c_{\alpha n})$. (We leave it as an exercise for the reader to show that the system of Section 8.3.2 always has at least one solution, given that $x_{\alpha 1} = 0$ for all α.)

DEFINITION **8.6**

For an expression α and an associated system of equations in Section 8.3.2 over a program graph, we define a *maximal solution set* $\{\hat{y}_{\alpha b}\}$ as a solution to the system of equations such that if $\{y'_{\alpha b}\}$ is any other solution set, we have $\hat{y}_{\alpha b} \geq y'_{\alpha b}$ for each node b.

The maximal solution set is of interest from the optimization point of view since it shows as available on exit from a block b (and, a fortiori, available on entrance to a block b') those expressions and only those expressions which are actually available on exit (entrance) to the block.

8.5. OBTAINING THE MAXIMAL SOLUTION SET

Next we shall consider the problem of algorithmically determining the maximal solution set. Let $G = (X, \Gamma)$ be the graph of a program P, and partition G into intervals. Order the nodes of each interval according to Algorithm 5.3. Select an arbitrary interval $I = I(h)$ and some expression α. For simplicity we shall solve the system of equations relating exclusively to α and designate the corresponding variables by block only: $y_{\alpha h} \leftrightarrow y_h$, $c_{\alpha h} \leftrightarrow c_h$, etc. Now

$$x_h = \prod_{p \in \Gamma^{-1}h} y_p = \Big(\prod_{p \in I \cap \Gamma^{-1}h} y_p\Big) \cdot \Big(\prod_{q \in \Gamma^{-1}h \sim I} y_q\Big) = y_I \cdot y_E$$

and

$$y_h = k_h \cdot x_h + c_h = k \cdot y_I \cdot y_E + c_h$$

where y_I is the contribution to x_h from the back-latches of the interval and y_E from the predecessors of h which are external to I.

The y_i for each $i \in I$ are partially dependent on the value of x_h, which is equal to either 0 or 1. If x_h were known, the values of the x_i and y_i could be completely determined since each $i \in \hat{\Gamma} h$ and $\Gamma^{-1}(I \sim \{h\}) \subset I$. Furthermore, if the values of the y_k are known for each $j \in I$ with $\Theta(j) < \Theta(i)$, then x_i and consequently y_i can be determined since, by Property 5.9, all the predecessors of i have lower-order numbers.

Thus the problem of finding a maximal solution for the redundancy equations of each interval reduces to finding the maximal solution x_h for its interval head. The influence of x_h within I continues until for each path $\mu[h, e_j] \subset I$, with $\Gamma e_j \not\subset I$, there is some $i \in \mu$ with either $c_i = 1$ or $k_i = 0$. Consequently, if $c_h = 1$ or $k_h = 0$, x_h is irrelevant to the entire interval (for the expression α, of course).

8.5.1. Dual Assumptions

The redundancy equations for I can be solved based on the dual assumptions $x_h^M = 1$ and $x_h^m = 0$. [In the interval $I(e)$, where e is the unique program entry node (which has no predecessors), we let $x_e^M = x_e^m = 0$, since nothing is available from the cleared store. The generality of the algorithm requires this assumption.] These assumptions immediately lead to dual tentative values x_j^M, x_j^m, and y_j^M, y_j^m for each node j in the interval. If there is a strongly connected component $S \subset I$, we recall that the set of back-latches $L = \{l \in I \mid h \in \Gamma l\}$ is nonempty. If for some $l \in L$ we have

$$y_l^M = 0 = y_l^m$$

it follows that $x_h = 0$, since $x_h = y_E y_I = y_E \cdot \prod_{l \in L} y_l = y_E \cdot 0 = 0$.

Superficially, there are four conceivable pairs of values for y_i^M together with y_i^m, for some arbitrary node $i \in I$.

CASE **8.7.I**

$y_i^M = 1$, $y_i^m = 1$. Here, either $c_i = 1$ or $x_i^M = x_i^m = 1$ and $k_i = 1$. The expression α is available on exit from block i regardless of its availability on entrance to h.

CASE **8.7.II**

$y_i^M = 0$, $y_i^m = 0$. Here, $c_i = 0$ and either $x_i^M = x_i^m = 0$ or $k_i = 0$. The expression α is not available on exit from block i, regardless of its availability on entrance to h.

CASE **8.7.III**

$y_i^M = 1$, $y_i^m = 0$. Here $c_i = 0$ and $k_i = 1$, since $y_i = k_i \prod_{p \in \Gamma^{-1} i} y_p + c_i$. Hence we must have $\prod_{p \in \Gamma^{-1} i} y_p^M = 1$ and $\prod_{p \in \Gamma^{-1} i} y_p^m = 0$. The expression α is available on exit from block i if and only if α is available on entrance to h.

CASE **8.7.IV**

$y_i^M = 0$, $y_i^m = 1$. Since c_i and k_i are a priori independent of x_h, this condition is impossible.

Thus there are only three possible cases, two of which are independent of the assumptions on x_h. Furthermore, if Case 8.7.II holds for any back-latch $l \in L$, we must then have that $x_h = x_h^m = 0$, so that the ambiguity of Case 8.7.III is completely resolved without reference to the value of y_E. In such a case, $y_i = y_i^m$ for each $i \in I$.

8.5.2. Passage to the Derived Graph Sequence

Next, consider the derived graph $(\mathcal{I}, \Gamma_{\mathcal{I}})$ whose vertices I are the intervals of G. Each arc (I_m, I_n) represents the arcs $(e_{m_1}, h_n), (e_{m_2}, h_n), \ldots, (e_{m_t}, h_n)$ in G, where the $e_{m_r} \in I_m$ and $I_n = I_n(h_n)$. Consequently, with each such arc (I_m, I_n), we shall associate the availability on exit variables

$$y^M(I_m, I_n) = \prod_{e_{m_r} \in I_m \cap \Gamma^{-1} h_n} y^M_{e_{m_r}}$$

and

$$y^m(I_m, I_n) = \prod_{e_{m_r} \in I_m \cap \Gamma^{-1} h_n} y^m_{e_{m_r}}$$

After $(\mathcal{I}, \Gamma_{\mathcal{I}})$ is partitioned into intervals, it may be possible to resolve the redundancy equations by examining the value of $x^M_{I_h}$ in the derived interval headed by I_h, for if $x^M_{I_h} = 0$, then for each I_q in that derived interval, $y^M(I_q, I_r)$ may be replaced by the value of $y^m(I_q, I_r)$ as a consequence of Case 8.7.II.

This process can be continued: $(\mathcal{I}^{(2)}, \Gamma_{\mathcal{I}^{(2)}})$, the derived graph of $(\mathcal{I}, \Gamma_{\mathcal{I}})$, can be analyzed; then $(\mathcal{I}^{(3)}, \Gamma_{\mathcal{I}^{(3)}})$ and so forth until the graphs converge to a single node. (The methods of Chapter 6 might be required for an irreducible graph.) Since G is a finite graph and each interval of G (and its associated derived graphs) has external predecessors for exactly one node, the process is convergent. Ergo, at one point in time the external entry conditions for each interval head will be known and the maximal solution established for every node in the program graph.

8.5.3. Solution of the Redundancy Equations by Computer

It is fortunate that the arithmetic units of most computers include *logical* instructions, which operate on the bit level over a machine word. The operators $+$ and $\cdot$ of the distributive lattice algebra coincide with *or* and *and*. The machine word can be thought of as a vector (called a *bit vector*) in which the ith component corresponds to a variable associated with the ith expression α in the system of redundancy equations. Thus with each block b in the program, we may associate a series of words C_b, K_b, Y_b^M, Y_b^m, X_b^M, and X_b^m which each contain a bit corresponding to each expression α. Each interval $I(h)$ can be ordered by the SNA, and starting with the dual availability on entrance assumptions for h, initial solutions Y_b^M, Y_b^m can be computed for each of the nodes of $I(h)$. [The SNA guarantees that to each of the immediate predecessors c of a node $b \in I(h)$ we shall have a value for Y_c^M, Y_c^m from which to compute X_b^M and X_b^m.] Next, the sequence of derived graphs is considered. The redundancy equations can be fully solved for the ordered nodes in the first interval $\mathcal{I}(e)$, since the initial availability of every expression is known on entrance to the program. To each node $j \in I(e) \sim \{e\}$ we have

$$Y_j \longleftarrow Y_j^M \cdot X_j^M$$

Within each of the other intervals $\mathcal{I}(h)$ of the derived graphs, it is possible to partially resolve the redundancy equation for each ordered node $c \in I(b) \in \mathcal{I}(h) \sim I(h)$ since all external influences on those ambiguous expressions of Case 8.7.III are either internal to $\mathcal{I}(h)$ (in which case their availability on exit is determined by the SNA) or pass through $I(h)$ from outside $\mathcal{I}(h)$. The latter ambiguities will be resolved when $\mathcal{I}(h)$ becomes an internal node of some derived interval $\mathcal{I}^{(p)}(h)$. For the nodes

$$c \in I(b) \in \mathcal{I}(h) \sim I(h)$$

we substitute

$$Y_c^M \longleftarrow X_b^M \cdot Y_c^M + \tilde{X}_b^M \cdot Y_c^m$$

and

$$Y_c^m \longleftarrow X_b^m \cdot Y_c^M + \tilde{X}_b^m \cdot Y_c^m$$

where X_b^M and X_b^m are computed over all the immediate predecessors of b in the original graph. The reader may show that it is only necessary to modify Y_c^M, since the minimal solution is of no practical interest.

The process eventually terminates with the unique interval in the last derived graph. The methods of node splitting described in Chapter 6 will be necessary for the treatment of an irreducible graph.

8.6. A DETAILED EXAMPLE

Example 8.8

We now show in detail the solution of a system of redundancy equations for a rather large computer program. The program graph is shown in Fig. 8.3. The program consists of 55 basic blocks and 32 expressions.

In each basic block approximately six expressions are computed within the block and are not subsequently annihilated, while approximately eight expressions are killed within each block. The vectors COMP and KILL, corresponding to C and K, are given in Fig. 8.4.

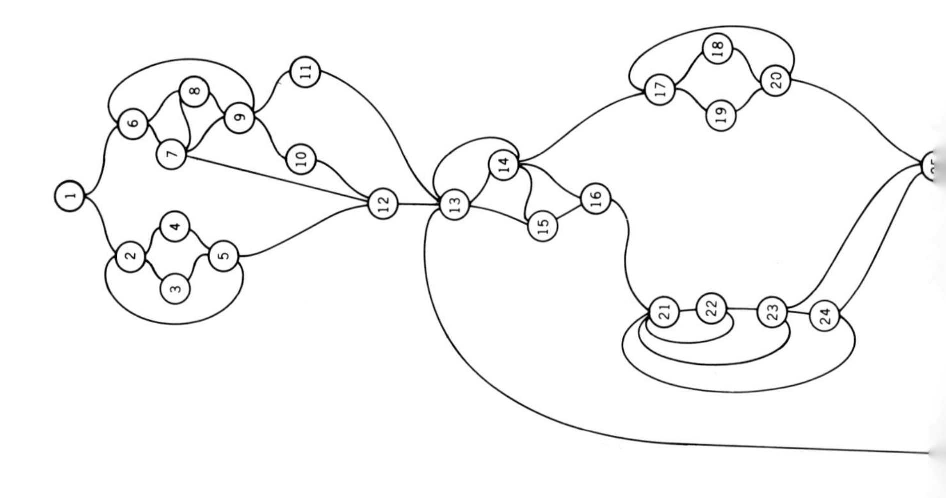
1
2
3
4
5
6
7
8
9
10
11
12
13
14
15
16
17
18
19
20
21
22
23
24

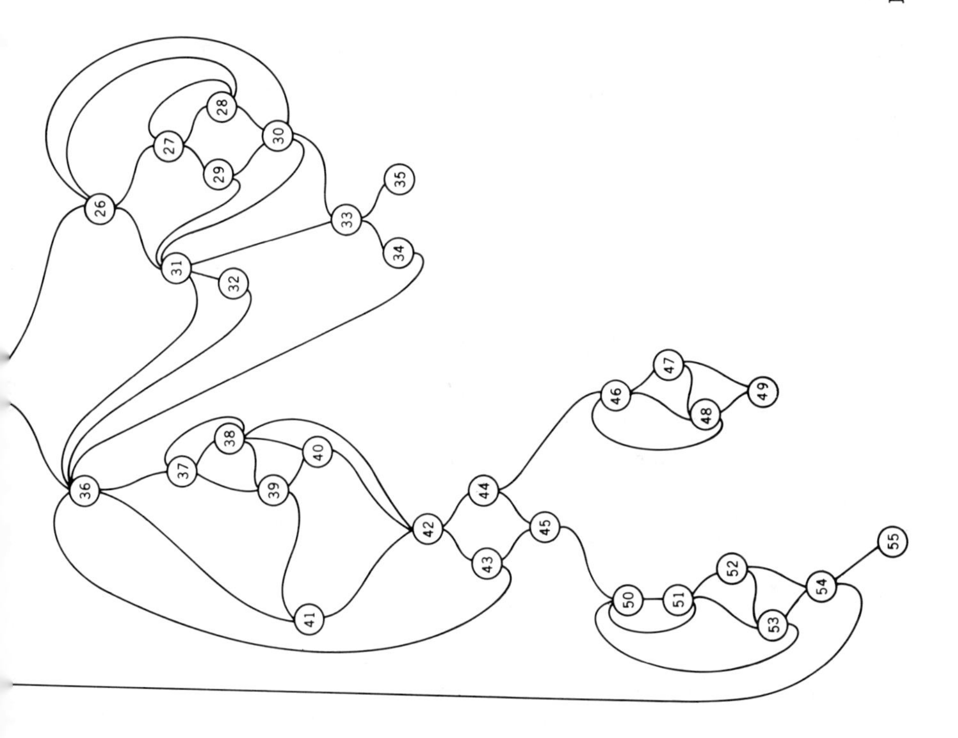
26
27
28
29
30
31
32
33
34
35
36
37
38
39
40
41
42
43
44
45
46
47
48
49
50
51
52
53
54
55

Figure 8.3

```
              COMP
 1|0010110000000001000000001000100100
 2|0101000000000000000000000010100100
 3|0100000100010001010000001011010 1
 4|0001011000001000000100000100101 1
 5|0001000000000100100010100010000 0
 6|0010010101000000011000000000100 0
 7|0000000001010100000100000000100 0
 8|0110000001010000010001100000010 00
 9|0000100001000001100101000000000 0
10|0000000000000001101000000000100 10
11|1010100000010100000000101000011 0
12|1001010100100110001000100001000 0
13|0000000001001001011000010000010 0
14|0000000000010000000000000001001010
15|0000001000000100000100100000100 0
16|0000100000000010000000000100000 1
17|1000100011100010100000000000000 10
18|0001000001011001010011000001100 1
19|0011000010000000101010001000000 0
20|0000000000010010000010011100110 0
21|0000010000000000000111000000100 0
22|0000001000100000100010000100110 1
23|1110000000100000000001000011010100
24|0100100001000010001001101110011 0
25|0100000000000000100000101010000 01
26|1000001010000000000101000000000 00
27|0000001100000100000000000100000 01
28|1000010000100000000100000011010 0
29|0000000010001010000010000001000 00
30|1100000100000000000101000110000 000
31|0000000000100000000000000100001101
32|0000110000010001000100010100000 00
33|0010100000000010000000010000000 00
34|0001000010000000000000000000010000
35|0001100100100000000000000000000 000
36|0001000000000110000000000000000 010
37|0010010101010000111000011000000 1
38|0101100001000010000110000001100 0
39|0000000001000001000001000000010 0
40|0110100000010000000000101011011 10
41|0100000000000000010000011000100 0
42|0010010001000011010010001100000 0
43|0000000010000000100101000000011 0
44|0000100001110001011000000000000 00
45|0001000001110110000100000011100 0
46|0000000000010000000100100000010 00
47|0000010001000000000001000000000 10
48|1100100000000001100010000111001 00
49|1110000000110110001000000000100 00
50|0000000010100010000100000000001 00
51|0100000000000100000001000100000 000
52|0000100001000010100000000100010 00
53|0000010010101011000011000000001 00
54|0101010001100010000000001000000 000
55|0010000100000000000000111101000100
```

```
              KILL
 1|1011111111111111011110110110110 00
 2|1011111111111101111111111111101111
 3|1011110101101110011111011101111 0
 4|0101110101001010101111011110111 1
 5|1101010011101011101011111111011 0
 6|1110110110011111111111111000111 11
 7|1110111111111101110011101111111 00
 8|0111101001111111111111101011110 11
 9|1011100111000110111111111011001 10
10|1111011110111111110111011101110 00
11|1011110011111110101111111110111 00
12|1011110111111010000011111111111 11
13|1111111111111111111110101111111 0111
14|1101010111111111111011011111111 101
15|1111011111110100111111111110111 101
16|1101111010111111010111011111011 110
17|1101101001110111010110101111111 11
18|0111011111101100101100010111011 1
19|1001110111111111111111101111110 111
20|1111110110111101100101111111111 11
21|1111111010110110101111110101111 1
22|1010111111111101001111101100111 01
23|1011011101111011101100110111011 1
24|1001111001110111011101111110001 10
25|1111111111111111101100111111111 0
26|1111111101111111111111011111100 11
27|1011111110111101111011001111111 0
28|1101000111111001111111111011011 00
29|1111101111000100111101001111011 1
30|1001111111111111111110111111001 0101
31|1110011111011111111011111111001 001
32|1111111101101110110011110011110 1
33|1111101111111010100111101001011 1
34|0000111111011111111011111111110 111
35|1010111110011111011111011011111 11
36|1111010111111101111111111110001 11
37|0110111100110101101111101111011 0
38|1111101111111111111101111111011 1111
39|1111111111010111101101111010110 1
40|1011111101111010111011101111110 11
41|1110111101101111011010111111011 1
42|1011111111111111101011011111010 11
43|1011111111111100101011111011111 11
44|1011100011010110101111111001011 10
45|1011101111100111110110111011111 1
46|1111001111111111110011111111011 111
47|1101111011100101101111000010001 1
48|1101011111110011001011011110111 0
49|1111010010111111010111111011101 1
50|0100111111101011101111110100011 11
51|1111101101111110111011111111011 111
52|1101111111100011111111011111011 1
53|0101100101111111111110111010011 10
54|1111101111111110100011000111111 1
55|1110110110111111011111001111001 1
```

Figure 8.4

It is readily seen that there are 18 intervals in the program graph. The graph has been ordered by Algorithm 5.3, and the intervals are

$$\begin{aligned}
I_1 &= \{1\} \\
I_2 &= \{2, 3, 4, 5\} \\
I_3 &= \{6, 7, 8, 9, 10, 11\} \\
I_4 &= \{12\} \\
I_5 &= \{13, 14, 15, 16\} \\
I_6 &= \{17, 18, 19, 20\} \\
I_7 &= \{21, 22, 23, 24\} \\
I_8 &= \{25\} \\
I_9 &= \{26\} \\
I_{10} &= \{27, 28, 29, 30\} \\
I_{11} &= \{31, 32\} \\
I_{12} &= \{33, 34, 35\} \\
I_{13} &= \{36\} \\
I_{14} &= \{37, 38, 39, 40\} \\
I_{15} &= \{41\} \\
I_{16} &= \{42, 43, 44, 45\} \\
I_{17} &= \{46, 47, 48, 49\} \\
I_{18} &= \{50, 51, 52, 53, 54, 55\}
\end{aligned}$$

To fully illustrate the method of solution, we shall detail the technique for the nodes of I_{18} and the first six program variables. We start with the dual assumptions $X_{50}^M = (1, 1, 1, 1, 1, 1)$ and $X_{50}^m = (0, 0, 0, 0, 0, 0)$. From these and the equations

$$Y_j = X_j \cdot \text{KILL}_j + \text{COMP}_j \qquad (50 \leq j \leq 55)$$

we readily obtain the tentative solution set shown in Fig. 8.5.

Since we can only have $Y_{\alpha n}^M = Y_{\alpha n}^m$ or $Y_{\alpha n}^M = 1$ while $Y_{\alpha n}^m = 0$, these results can be summarized as

$$\begin{aligned}
Y_{50} &= (0, \neq, 0, 0, \neq, \neq) \\
Y_{51} &= (0, \ \ 1, 0, 0, \neq, \ \ 0) \\
Y_{52} &= (0, \ \ 1, 0, 0, \ \ 1, \ \ 0) \\
Y_{53} &= (0, \ \ 1, 0, 0, \neq, \ \ 1)
\end{aligned}$$

$$X^M_{50} = (1, 1, 1, 1, 1, 1)$$
$$Y^M_{50} = (0, 1, 0, 0, 1, 1)$$
$$X^M_{51} = (0, 1, 0, 0, 1, 1)$$
$$Y^M_{51} = (0, 1, 0, 0, 1, 0)$$
$$X^M_{52} = (0, 1, 0, 0, 1, 0)$$
$$Y^M_{52} = (0, 1, 0, 0, 1, 0)$$
$$X^M_{53} = (0, 1, 0, 0, 1, 0)$$
$$Y^m_{53} = (0, 1, 0, 0, 1, 1)$$
$$X^M_{54} = (0, 1, 0, 0, 1, 0)$$
$$Y^m_{54} = (0, 1, 0, 1, 1, 1)$$
$$X^M_{55} = (0, 1, 0, 1, 1, 1)$$
$$Y^m_{55} = (0, 1, 1, 0, 1, 1)$$

$$X^m_{50} = (0, 0, 0, 0, 0, 0)$$
$$Y^m_{50} = (0, 0, 0, 0, 0, 0)$$
$$X^m_{51} = (0, 0, 0, 0, 0, 0)$$
$$Y^m_{51} = (0, 1, 0, 0, 0, 0)$$
$$X^m_{52} = (0, 1, 0, 0, 0, 0)$$
$$Y^m_{52} = (0, 1, 0, 0, 1, 0)$$
$$X^m_{53} = (0, 1, 0, 0, 0, 0)$$
$$Y^m_{53} = (0, 1, 0, 0, 0, 1)$$
$$X^m_{54} = (0, 1, 0, 0, 0, 0)$$
$$Y^m_{54} = (0, 1, 0, 1, 0, 1)$$
$$X^m_{55} = (0, 1, 0, 1, 0, 1)$$
$$Y^m_{55} = (0, 1, 1, 0, 0, 1)$$

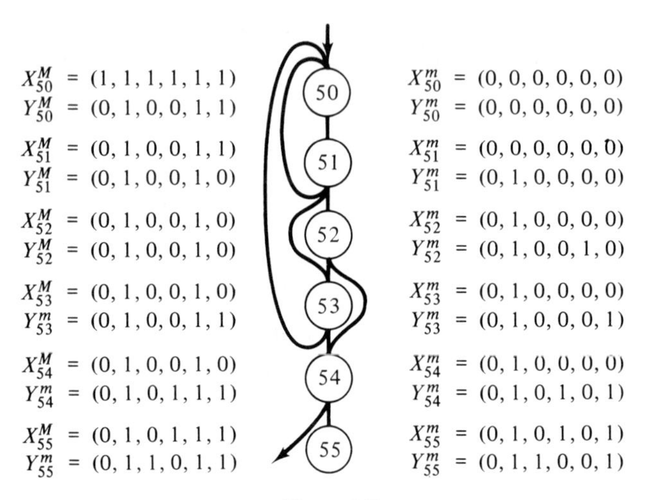

Figure 8.5

$$Y_{54} = (0, \quad 1, 0, 1, \neq, \quad 1)$$
$$Y_{55} = (0, \quad 1, 1, 0, \neq, \quad 1)$$

where $\neq$ represents the case $Y^M_{\alpha n} > Y^m_{\alpha n}$.

The first-order results for the entire program graph are summarized in Fig. 8.6.

```
INTERVAL 1
EXIT  1:  00101100000000100000001000100100
INTERVAL 2
EXIT  2:  ≠1≠1≠≠≠≠≠≠≠≠≠0≠≠≠≠≠≠≠≠≠≠1≠10≠1≠≠
EXIT  3:  ≠1≠1≠≠010≠≠1≠0≠101≠≠≠≠0≠1≠11≠1≠1
EXIT  4:  0101≠11≠0≠0010≠0≠0≠1≠≠0≠11101111
EXIT  5:  01010≠000≠00≠1≠010≠01≠1≠1≠1001≠0
INTERVAL 3
EXIT  6:  ≠≠10≠101≠10≠≠≠≠≠≠11≠≠≠≠≠000≠1≠≠≠
EXIT  7:  ≠≠10≠101≠101≠1≠≠≠001≠≠0≠000≠1≠00
EXIT  8:  0110≠000111≠≠≠≠≠100≠110≠000≠1000
EXIT  9:  00101000≠1000≠≠11001≠10≠00000000
EXIT 10:  00100000≠0000≠111101≠00≠00010010
EXIT 11:  10101000≠10101010001≠11≠10000110
INTERVAL 4
EXIT 12:  10≠1≠101≠≠1≠0110001≠≠≠1≠≠≠≠1≠≠≠≠
INTERVAL 5
EXIT 13:  ≠≠≠≠≠≠≠≠≠1≠≠1≠≠1≠11≠0≠≠1≠≠≠≠01≠≠
EXIT 14:  ≠≠0≠0≠0≠≠11≠1≠≠1≠01≠0≠≠1≠1≠≠111≠
EXIT 15:  ≠≠0≠0≠1≠≠1≠01101≠0110≠11≠0≠≠110≠
EXIT 16:  ≠≠0≠1≠00≠0≠01≠10≠01≠00≠1≠10≠1101
```

Figure 8.6

```
INTERVAL 6
EXIT 17:  1≠0≠10≠0111≠0≠1≠1≠0≠≠0≠0≠≠≠≠≠≠1≠
EXIT 18:  0≠0100≠011111≠01110≠11000≠≠11≠11
EXIT 19:  10111000111≠0≠1≠1≠1≠10≠01≠≠0≠≠1≠
EXIT 20:  00010000101 10≠1≠100≠100111≠0111≠
INTERVAL 7
EXIT 21:  ≠≠≠≠≠1≠0≠0≠≠0≠≠0≠0≠111≠≠0≠0≠1≠≠≠
EXIT 22:  ≠0≠0≠110≠01≠00≠010≠1110≠010≠1101
EXIT 23:  1110011001 1≠00≠010≠1000≠11010101
EXIT 24:  1100111001 1≠0010001101 1≠11100110
INTERVAL 8
EXIT 25:  ≠1≠≠≠≠≠≠≠≠≠≠≠≠≠≠1≠0≠≠01≠1≠1≠≠≠≠1
INTERVAL 9
EXIT 26:  1≠≠≠≠1≠1≠≠≠≠≠≠≠≠≠≠1≠10≠≠≠≠≠≠00≠≠
INTERVAL 10
EXIT 27:  ≠0≠≠≠11≠0≠≠≠10≠≠≠≠0≠≠001≠≠≠≠≠≠1
EXIT 28:  100≠0101≠01≠≠00≠≠≠≠1≠≠000≠11≠100
EXIT 29:  ≠0≠≠≠0111000111 0≠≠≠01≠001≠1≠0≠≠1
EXIT 30:  110≠0011≠000≠000≠1≠1≠≠01100≠0≠00
INTERVAL 11
EXIT 31:  ≠≠≠00≠≠≠≠≠1≠≠≠≠≠≠≠0≠≠≠≠1≠≠001101
EXIT 32:  ≠≠≠011≠≠0≠11≠≠≠1≠≠01≠≠≠101001101
INTERVAL 12
EXIT 33:  ≠≠1≠10≠≠≠≠≠≠≠≠1≠0≠00≠≠≠10≠00≠0≠≠≠
EXIT 34:  000110≠≠1≠0≠≠1≠0≠00≠≠≠10≠0010≠≠≠
EXIT 35:  ≠01110≠1≠01≠≠1≠0000≠≠≠00≠00≠0≠≠≠
INTERVAL 13
EXIT 36:  ≠≠≠10≠0≠≠≠≠≠110≠≠≠≠≠≠≠≠≠≠≠≠000≠1≠
INTERVAL 14
EXIT 37:  0≠10≠1≠101≠10≠0≠111≠≠≠≠11≠≠≠0≠≠1
EXIT 38:  011110≠101≠10≠1≠11011≠≠110≠11≠≠1
EXIT 39:  0≠10≠0≠101010≠01100≠01≠110≠00101
EXIT 40:  011010≠1010100001000 01≠111101111
INTERVAL 15
EXIT 41:  ≠1≠0≠≠≠≠0≠≠0≠≠≠≠01≠0≠0≠11≠≠≠1≠≠≠
INTERVAL 16
EXIT 42:  ≠01≠≠1≠≠≠1≠≠≠≠11010≠10≠≠11≠0≠0≠≠
EXIT 43:  ≠01≠≠1≠≠11≠≠≠001110111≠≠01≠0≠11≠
EXIT 44:  ≠01≠1000≠1110≠11011≠10≠≠00≠0≠0≠0
EXIT 45:  ≠011≠000≠11101110101 10≠≠001110≠0
INTERVAL 17
EXIT 46:  ≠≠≠≠00≠≠≠≠≠1≠≠≠≠≠01≠≠1≠≠≠≠0≠1≠≠≠
EXIT 47:  ≠≠0≠01≠0≠1≠00≠0≠≠01≠1100000000 1≠
EXIT 48:  110≠10≠0≠≠≠00011001 1≠100111001≠0
EXIT 49:  111≠0000≠011011≠001≠≠1000001 00≠0
INTERVAL 18
EXIT 50:  0≠00≠≠≠≠1≠10≠01≠≠0≠1≠≠≠0≠000≠1≠≠
EXIT 51:  0100≠0≠≠0≠10100≠≠001≠≠≠1≠000≠1≠≠
EXIT 52:  010010≠≠0110001≠1001≠≠01100011≠≠
EXIT 53:  0100≠10≠1≠101011≠0011101 0000≠1≠0
EXIT 54:  0101≠10≠01100010≠000≠≠010000≠1≠0
EXIT 55:  0110≠101001000100000111101000 1≠0
```

Figure 8.6 (*cont.*)

The first derived graph consists of 18 nodes, and partitions into the intervals

$$\mathcal{I}_1 = \{I_1, I_2, I_3, I_4\}$$
$$\mathcal{I}_2 = \{I_5, I_6, I_7, I_8\}$$
$$\mathcal{I}_3 = \{I_9, I_{10}, I_{11}, I_{12}\}$$
$$\mathcal{I}_4 = \{I_{13}, I_{14}, I_{15}, I_{16}, I_{17}, I_{18}\}$$

The graph is shown in Fig. 8.7.

Consider $\mathcal{I}_3$ an isolated entity. We start by ignoring I_9 (since the outside influences are not yet known) and proceed to compute $X^M_{I_{10}}$ and $X^m_{I_{10}}$. Since $\Gamma^{-1}I_{10} = \Gamma^{-1}\{27\} = \{26, 28\}$, $X_{I_{10}} = Y_{26} \cdot Y_{28}$. The previous results yield

$$X^M_{I_{10}} = (1, 1, 1, 1, 1, 1) \cdot (1, 0, 0, 1, 0, 1) = (1, 0, 0, 1, 0, 1)$$
$$X^m_{I_{10}} = (1, 0, 0, 0, 0, 0) \cdot (1, 0, 0, 0, 0, 1) = (1, 0, 0, 0, 0, 0)$$

This information permits the computation of Y^M_j and Y^m_j $(27 \leq j \leq 35)$; e.g.,

$$\begin{aligned} Y^M_{27} &= (1, 0, 0, 1, 0, 1) \cdot (1, 0, 1, 1, 1, 1) + (0, 1, 1, 0, 1, 0) \cdot (0, 0, 0, 0, 0, 0) \\ &= (1, 0, 0, 1, 0, 1) \end{aligned}$$
$$\begin{aligned} Y^m_{27} &= (1, 0, 0, 0, 0, 0) \cdot (1, 0, 1, 1, 1, 1) + (0, 1, 1, 1, 1, 1) \cdot (0, 0, 0, 0, 0, 0) \\ &= (1, 0, 0, 0, 0, 0) \end{aligned}$$

and

$$\begin{aligned} Y^M_{28} &= (1, 0, 0, 1, 0, 1) \qquad & Y^m_{28} &= (1, 0, 0, 0, 0, 1) \\ Y^M_{29} &= (1, 0, 0, 1, 0, 0) \qquad & Y^m_{29} &= (1, 0, 0, 0, 0, 0) \\ Y^M_{30} &= (1, 1, 0, 1, 0, 0) \qquad & Y^m_{30} &= (1, 1, 0, 0, 0, 0) \end{aligned}$$

Similarly, $\Gamma^{-1}I_{11} = \Gamma^{-1}\{31\} = \{26, 29, 30\}$. Thus $X_{I_{11}} = Y_{26} \cdot Y_{29} \cdot Y_{30}$:

$$X^M_{I_{11}} = (1, 1, 1, 1, 1, 1) \cdot (1, 0, 0, 1, 0, 0) \cdot (1, 1, 0, 1, 0, 0) = (1, 0, 0, 1, 0, 0)$$
$$X^m_{I_{11}} = (1, 0, 0, 0, 0, 0) \cdot (1, 0, 0, 0, 0, 0) \cdot (1, 1, 0, 0, 0, 0) = (1, 0, 0, 0, 0, 0)$$

Thus

$$\begin{aligned} Y^M_{31} &= (1, 0, 0, 0, 0, 0) \qquad & Y^m_{31} &= (1, 0, 0, 0, 0, 0) \\ Y^M_{32} &= (1, 0, 0, 0, 1, 1) \qquad & Y^m_{32} &= (1, 0, 0, 0, 1, 1) \end{aligned}$$

Since $\Gamma^{-1}I_{12} = \Gamma^{-1}\{33\} = \{30, 31\}$, we obtain

$$X^M_{I_{12}} = (1, 0, 0, 0, 0, 0) = X^m_{I_{12}}$$

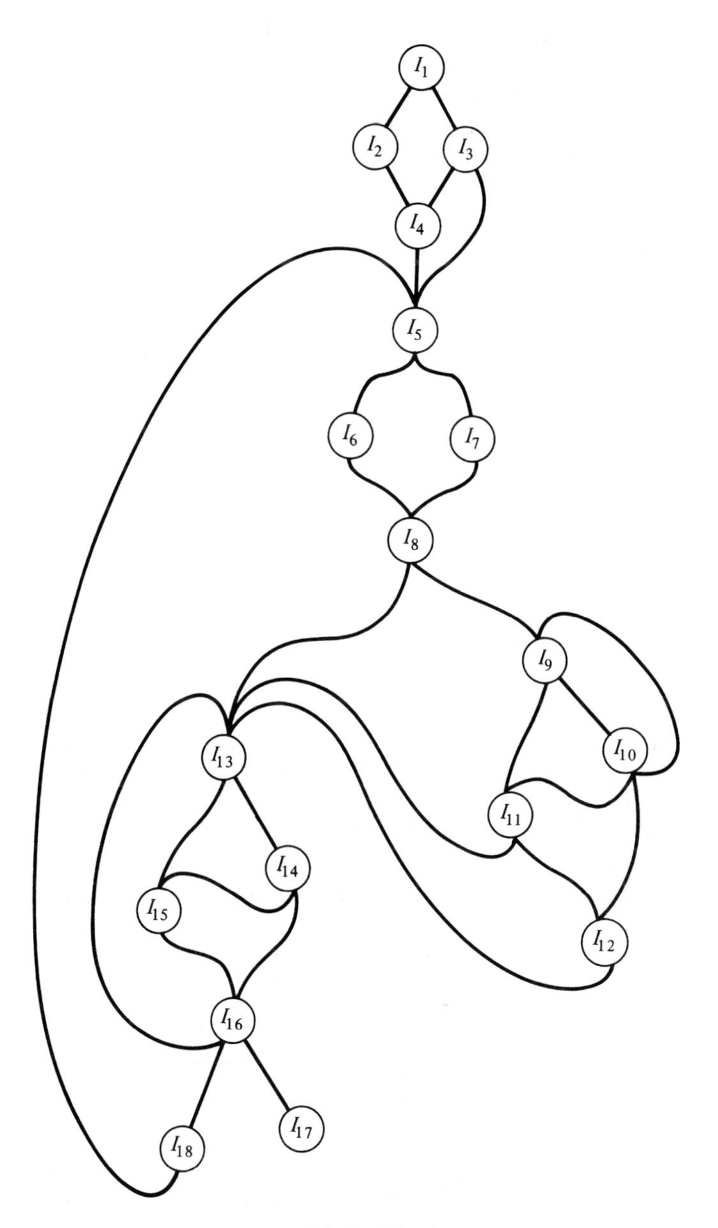

Figure 8.7

and

$$\begin{aligned} Y_{33}^{M} &= (1, 0, 1, 0, 1, 0) \qquad & Y_{33}^{m} &= (1, 0, 1, 0, 1, 0) \\ Y_{34}^{M} &= (0, 0, 0, 1, 1, 0) \qquad & Y_{34}^{m} &= (0, 0, 0, 1, 1, 0) \\ Y_{35}^{M} &= (1, 0, 1, 1, 1, 0) \qquad & Y_{35}^{m} &= (1, 0, 1, 1, 1, 0) \end{aligned}$$

Employing the symbol Δ to indicate that the new solution for one of the $Y_{\alpha j}$ differs from the previous solution, we summarize with

$$\begin{aligned} Y_{27} &= (1, 0, 0, \neq, 0, \neq)\Delta \\ Y_{28} &= (1, 0, 0, \neq, 0, 1) \\ Y_{29} &= (1, 0, 0, \neq, 0, 0)\Delta \\ Y_{30} &= (1, 1, 0, \neq, 0, 0) \\ Y_{31} &= (1, 0, 0, 0, 0, 0)\Delta \\ Y_{32} &= (1, 0, 0, 1, 1, 1)\Delta \\ Y_{33} &= (1, 0, 1, 0, 1, 0)\Delta \\ Y_{34} &= (0, 0, 0, 1, 1, 0) \\ Y_{35} &= (1, 0, 1, 1, 1, 0)\Delta \end{aligned}$$

The results for the entire graph at this stage of the analysis are given in Fig. 8.8.

```
INTERVAL 1
EXIT  1 : 00101100000000100000001000100100
EXIT  2 : 01010100000000100000001010100100Δ
EXIT  3 : 01010101000100110100000010110101Δ
EXIT  4 : 01010110000010100001000011101111Δ
EXIT  5 : 01010100000001101000101010100100Δ
EXIT  6 : 00101101010000100110000000001000Δ
EXIT  7 : 00101101010101100001000000001000Δ
EXIT  8 : 01101000111000101000110000001000Δ
EXIT  9 : 00101000010000111001010000000000Δ
EXIT 10 : 00100000000000111101000000010010Δ
EXIT 11 : 10101000010101010001011010000110Δ
EXIT 12 : 10010101001001100010001000010000Δ
INTERVAL 2
EXIT 13 : ≠≠≠≠≠≠≠≠≠1≠≠1≠≠1≠11≠0≠≠1≠≠≠≠01≠≠
EXIT 14 : ≠≠0≠0≠0≠≠11≠1≠≠1≠01≠0≠≠1≠1≠≠111≠
EXIT 15 : ≠≠0≠0≠1≠≠1≠01101≠0110≠11≠0≠≠110≠
EXIT 16 : ≠≠0≠1≠00≠0≠01≠10≠01≠00≠1≠10≠1101
EXIT 17 : 100≠1000111≠0≠1≠100≠0000≠1≠0111≠Δ
EXIT 18 : 0001000011111≠01110≠110001≠11111Δ
EXIT 19 : 10111000111≠0≠1≠101≠100011≠0111≠Δ
EXIT 20 : 0001000010110≠1≠100≠100111≠0111≠
```

Figure 8.8

```
EXIT 21 : ≠000010000≠000≠000≠1110≠01001100Δ
EXIT 22 : ≠0000110001000≠010≠1110≠01001101Δ
EXIT 23 : 11100110011000≠010≠1000≠11010101Δ
EXIT 24 : 11001110011000100011011≠11100110Δ
EXIT 25 : 01000000001000≠1000≠010111000101Δ
INTERVAL 3
EXIT 26 : 1≠≠≠≠≠1≠1≠≠≠≠≠≠≠≠1≠10≠≠≠≠≠≠00≠≠
EXIT 27 : 100≠0≠11≠0≠≠≠10≠≠≠≠0≠0001≠≠≠0001Δ
EXIT 28 : 100≠0101≠01≠≠00≠≠≠≠1≠0000≠110100Δ
EXIT 29 : 100≠001110001110≠≠≠010001≠1≠0001Δ
EXIT 30 : 110≠0011≠000≠000≠1≠1≠001100≠0000Δ
EXIT 31 : 1000001≠≠010≠000≠≠00≠001≠0001101Δ
EXIT 32 : 1000111≠0011≠001≠≠01≠00101001101Δ
EXIT 33 : 1010101≠≠000≠100≠000≠010≠00000000Δ
EXIT 34 : 0001101≠1000≠100≠000≠010≠00100000Δ
EXIT 35 : 10111011≠010≠1000000≠000≠00000000Δ
INTERVAL 4
EXIT 36 : ≠≠≠10≠0≠≠≠≠≠110≠≠≠≠≠≠≠≠≠≠≠000≠1≠
EXIT 37 : 0≠10010101≠10≠0≠111≠≠≠≠110000≠≠1Δ
EXIT 38 : 0111100101≠10≠1≠11011≠≠110011≠≠1Δ
EXIT 39 : 0≠10000101010≠01100≠01≠110000101Δ
EXIT 40 : 0110100101010000100001≠111101111Δ
EXIT 41 : 01≠0000≠0≠000≠0≠010000≠110001≠0≠Δ
EXIT 42 : 0010010≠01000011010010≠11100100≠Δ
EXIT 43 : 0010010≠11000001110111≠10100111≠Δ
EXIT 44 : 0010100001110011011010≠100001000Δ
EXIT 45 : 0011000001110111010110≠100111000Δ
EXIT 46 : 000000000≠≠100110010≠10000001000Δ
EXIT 47 : 0000010001≠0000100101100000000010Δ
EXIT 48 : 110010000≠≠000110011≠10011100100Δ
EXIT 49 : 11100000001101110010≠10000010000Δ
EXIT 50 : 000000001≠1000100001≠00000000≠100Δ
EXIT 51 : 010000000≠1010000001≠0010000≠100Δ
EXIT 52 : 01001000011000101001≠00110001100Δ
EXIT 53 : 010001001≠101011000111010000≠100Δ
EXIT 54 : 01010100011000100000≠0010000≠100Δ
EXIT 55 : 01100101001000100000111101000100Δ
```

Figure 8.8 (*cont.*)

It should be noted that the redundancy equations for $\mathcal{I}_1$ have been completely solved since there is complete agreement between the Y_i^M and the Y_i^m, $1 \leq i \leq 12$. Since $\mathcal{I}_2 \cap \Gamma^{-1} I_5 = \varnothing$, the redundancy equations for the nodes of I_5 are not affected by this stage of the processing, while virtually every other redundancy equation for $\mathcal{I}_2$ has been partially resolved.

The second derived graph is partitioned into two intervals:

$$\mathcal{I}_1^{(2)} = \{\mathcal{I}_1\}$$

$$\mathcal{I}_2^{(2)} = \{\mathcal{I}_2, \mathcal{I}_3, \mathcal{I}_4\}$$

as shown in Fig. 8.9.

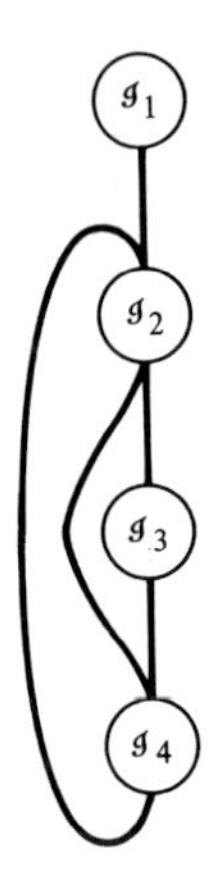

Figure 8.9

Application of the technique used on the first derived graph yields the results shown in Fig. 8.10. Note that the redundancy equations have been solved for all of the graph except $\mathcal{G}_2$ and node $\{26\}$.

```
INTERVAL 1
EXIT  1 : 0010110000000001000000001000100100
EXIT  2 : 0101010000000001000000001010100100
EXIT  3 : 010101010001001101000000010110101
EXIT  4 : 0101011000000101000010000011101111
EXIT  5 : 01010100000001101000101010100100
EXIT  6 : 00101101010000100110000000001000
EXIT  7 : 001011010101011000010000000001000
EXIT  8 : 011010001110001010001100000001000
EXIT  9 : 00101000010000111001010000000000
EXIT 10 : 00100000000000111101000000010010
EXIT 11 : 10101000010101010001011010000110
EXIT 12 : 10010101001001100010001000010000
INTERVAL 2
EXIT 13 : ≠≠≠≠≠≠≠≠≠1≠≠1≠≠1≠11≠0≠≠1≠≠≠≠01≠≠
EXIT 14 : ≠≠0≠0≠0≠≠11≠1≠≠1≠01≠0≠≠1≠1≠≠111≠
EXIT 15 : ≠≠0≠0≠1≠≠1≠01101≠0110≠11≠0≠≠110≠
EXIT 16 : ≠≠0≠1≠00≠0≠01≠10≠01≠00≠1≠10≠1101
EXIT 17 : 100≠1000111≠0≠1≠100≠0000≠1≠0111≠
EXIT 18 : 0001000011111≠01110≠110001≠11111
EXIT 19 : 10111000111≠0≠1≠101≠100011≠0111≠
EXIT 20 : 0001000010110≠1≠100≠100111≠0111≠
EXIT 21 : ≠000010000≠000≠000≠1110≠01001100
EXIT 22 : ≠0000110001000≠010≠1110≠01001101
EXIT 23 : 11100110011000≠010≠1000≠11010101
EXIT 24 : 11001110011000100011011≠11100110
EXIT 25 : 0100000000001000≠1000≠010111000101
EXIT 26 : 10000010100000000001≠100000000000Δ
EXIT 27 : 10000011000001000000000010000001Δ
EXIT 28 : 10000101001000000001000000110100Δ
EXIT 29 : 10000011100011100000100010100001Δ
EXIT 30 : 11000011000000000101000110000000Δ
EXIT 31 : 10000010001000000000000100001101Δ
EXIT 32 : 10001110001100010001000101001101Δ
```

Figure 8.10

```
EXIT 33 : 10101010000001000000001000000000Δ
EXIT 34 : 00011010100001000000001000010000Δ
EXIT 35 : 10111011001001000000000000000000Δ
EXIT 36 : 00010000000001100000000000000010Δ
EXIT 37 : 00100101010100001110000110000001Δ
EXIT 38 : 01111001010100101101100110011001Δ
EXIT 39 : 00100001010100011000010110000101Δ
EXIT 40 : 01101001010100001000010111101111Δ
EXIT 41 : 01000000000000000100000110001000Δ
EXIT 42 : 00100100010000110100100111001000Δ
EXIT 43 : 00100100110000011101110101001110Δ
EXIT 44 : 00101000011100110110100100001000Δ
EXIT 45 : 00110000011101110101100100111000Δ
EXIT 46 : 00000000000100110010010000001000Δ
EXIT 47 : 00000100010000010010110000000010Δ
EXIT 48 : 11001000000000110011010011100100Δ
EXIT 49 : 11100000001101110010010000010000Δ
EXIT 50 : 00000000101000100001000000000100Δ
EXIT 51 : 01000000001010000001000100000100Δ
EXIT 52 : 01001000011000101001000110001100Δ
EXIT 53 : 01000100101010110001110100000100Δ
EXIT 54 : 01010100011000100000000100000100Δ
EXIT 55 : 01100101001000100000111101000100
```

Figure 8.10 (*cont.*)

The third derived graph has one interval and is shown in Fig. 8.11, and the complete solution of the redundancy equations appears in Fig. 8.12.

Figure 8.11

```
INTERVAL 1
EXIT  1 : 00101100000000100000001000100100
EXIT  2 : 01010100000000100000001010100100
EXIT  3 : 01010101000100110100000010110101
EXIT  4 : 01010110000010100001000011101111
EXIT  5 : 01010100000001101000101010100100
EXIT  6 : 00101101010000100110000000001000
EXIT  7 : 00101101010101100001000000001000
EXIT  8 : 01101000111000101000110000001000
EXIT  9 : 00101000010000111001010000000000
EXIT 10 : 00100000000000111101000000010010
EXIT 11 : 10101000010101010001011010000110
EXIT 12 : 10010101001001100010001000010000
EXIT 13 : 00000000010010010110000100000100Δ
EXIT 14 : 00000000011010010010000101001110Δ
EXIT 15 : 00000010010011010011001100001100Δ
EXIT 16 : 00001000000010100010000101001101Δ
EXIT 17 : 10001000111000101000000001001110Δ
EXIT 18 : 00010000111110011100110001011111Δ
```

Figure 8.12 (*continued on next page*)

```
EXIT 19 : 10111000111000101010100011001110Δ
EXIT 20 : 00010000101100101000100111001110Δ
EXIT 21 : 00000100000000000001110001001100Δ
EXIT 22 : 00000110001000001001110001001101Δ
EXIT 23 : 11100110011000001001000011010101Δ
EXIT 24 : 11001110011000100011011011100110Δ
EXIT 25 : 01000000001000010000010111000101Δ
EXIT 26 : 10000010100000000010100000000000Δ
EXIT 27 : 10000011000001000000000010000001
EXIT 28 : 10000101001000000001000000110100
EXIT 29 : 10000011100011100000100010100001
EXIT 30 : 11000011000000000101000110000000
EXIT 31 : 10000010001000000000000100001101
EXIT 32 : 10001110001100010001000101001101
EXIT 33 : 10101010000001000000001000000000
EXIT 34 : 00011010100001000000001000010000
EXIT 35 : 10111011001001000000000000000000
EXIT 36 : 00010000000011000000000000000010
EXIT 37 : 00100101010100001110000110000001
EXIT 38 : 01111001010100101101100110011001
EXIT 39 : 00100001010100110000010110000101
EXIT 40 : 01101001010100001000010111101111
EXIT 41 : 01000000000000001000000110001000
EXIT 42 : 00100100010000110100100111001000
EXIT 43 : 00100100110000011101110101001110
EXIT 44 : 00101000011100110110100100001000
EXIT 45 : 00110000011101110101100100111000
EXIT 46 : 00000000000100110010010000001000
EXIT 47 : 00000100010000010010110000000010
EXIT 48 : 11001000000000110011010011100100
EXIT 49 : 11100000001101110010010000010000
EXIT 50 : 00000000101000100001000000000100
EXIT 51 : 01000000001010000001000100000100
EXIT 52 : 01001000011000101001000110001100
EXIT 53 : 01000100101010110001110100000100
EXIT 54 : 01010100011000100000000100000100
EXIT 55 : 01100101001000100000111101000100
```

Figure 8.12 (*cont.*)

EXERCISE

1. Assume that the nodes of the entire graph have been ordered by the modification to the SNA indicated in the exercise to Chapter 5. Consider the system of equations

$$y_{\alpha b} = k_{\alpha b} x_{\alpha b} + c_{\alpha b}$$

$$x_{\alpha b} = \prod_{\substack{p \in \Gamma^{-1} b \\ \Theta(p) < \Theta(b)}} y_{\alpha p}$$

Show that this system of equations can be solved directly, by substituting values according to the order of the nodes.

Show that the solution obtained in this manner is an upper bound to the maximal solution to the equations of Section 8.3.2, which is bounded above by the $y^M_{\alpha b}$.

Given the solution to the redundancy equations of this chapter, what further analysis is necessary in order that an expression α in b be redundant when $x_{\alpha b} = 1$?

2. Consider the following code sequence from some basic block:

$$\begin{aligned}
c &\longleftarrow a + b \\
x &\longleftarrow a \\
d &\longleftarrow x + b \\
y &\longleftarrow b \\
e &\longleftarrow x + y \\
a &\longleftarrow x \\
f &\longleftarrow a + b \\
b &\longleftarrow d \\
g &\longleftarrow a + b \\
h &\longleftarrow a + y
\end{aligned}$$

Clearly, the initial value of $a + b$ is being stored into each of c, d, e, and f, while g receives a different value. The method of α-expressions does not reveal all of the formal identities in such code sequences. However, one may assign *value numbers* to each variable and, by replacing each argument of an α-expression with its current value number, cause the identities to become apparent.

In this case, we would have

$$\begin{aligned}
a &\longleftrightarrow \mathbf{11} \\
b &\longleftrightarrow \mathbf{12} \\
a + b &\longleftrightarrow \mathbf{11} + \mathbf{12} : \mathbf{13} \\
c &\longleftrightarrow \mathbf{13}
\end{aligned}$$

i.e., the α-expression $a + b$ becomes the value number expression **11** + **12**. This expression is seen to be unique, so it is given the value number **13** which is the current value number associated with the variable c. After $x \leftarrow a$ gives x the value number **11**, we see that $x + b$ becomes **11** + **12**, i.e., **13**.

Now, $b \leftarrow d$ gives b the new value **13**, so that $a + b$ becomes **11** + **13**. This is a new α-expression, so it receives a fresh value number, **14**. But $a + y$ is still value numbered as **13**, so all formal identities have been exposed.

a. Show that this method can be used to expose some redundant subexpressions between blocks in an acyclic graph. Produce a simple example of an unrecognizable redundant subexpression in an acyclic graph.

b. What modifications must be made in the case of a graph with cycles? [*Hint*. Consider a graph with nodes m and n such that $n \in \mathbf{\Gamma} m$, $\{m, n\} = \mathbf{\Gamma}^{-1} n$, and where $a + b$ is computed downward in m, computed as the first instruction of n, and then killed in n.]

9 CONSTANT SUBSUMPTION, COMMON SUBEXPRESSION SUPPRESSION, AND CODE MOTION

9.1. MACHINE-INDEPENDENT OPTIMIZATIONS

Optimization involves the reduction of certain cost functions associated with a computer program. These cost functions include *space*, the amount of machine store occupied by the program and its data, and *time*. In our context we restrict the notion of time to cover just the period in which the program is running on a computer, rather than that required to encode the program or for it to be compiled and analyzed.

In this chapter we shall introduce a number of techniques which effect one or both of these optimizations on code. These techniques may be called *machine-independent optimizations*, since they tend to reduce the cost functions, to some extent, on all computers. In subsequent chapters, techniques will be introduced that do not have this property but which depend very strongly on the class of computer the program is to be executed on, if not the precise computer model selected.

9.2. COMPILE-TIME OPTIMIZATIONS

We need to distinguish between two classes of computations: those that necessarily occur when a program is executing, and those that may be performed prior to program execution, say during the program's compilation. These are, respectively, called *run-time computations* and *compile-time computations*.

There are obvious advantages to identifying and performing all of a program's compile-time computations with the compiler. Indeed, doing so eliminates the necessity to produce code for them, a space optimization, and, if the program is to be executed more than once, a clear saving in computer time is realized.

9.2.1. Constant Propagation

Compile-time computations manifest themselves in varied forms. The simplest case is characterized by

$$\alpha \longleftarrow 32 \times 256$$

which can be replaced by

$$\alpha \longleftarrow 8192$$

Less obvious is

$$\pi \longleftarrow 3.14159$$
$$\vdots$$
$$c \longleftarrow 2 \times \pi \times r$$

which could be replaced by

$$\pi \longleftarrow 3.14159$$
$$\vdots$$
$$c \longleftarrow 6.28318 \times r$$

Not all compile-time computations occur directly in the programmer's code but are actually generated by the compiler, for example, the FORTRAN code

```
DIMENSION A(100,200)
. . .
X = A(K,25) + Y
```

FORTRAN arrays are stored in columnar order; i.e., the first column is stored with A(1,1) followed by A(2,1), . . . , followed by A(100,1), then A(1,2) A(2,2), A(3,2), etc. Thus the element A(I,J) occurs in the 200*(J−1) + I−1th location following A(1,1). Thus A(K,25) would be $24 \times 200 + K-1$ computer locations (words, bytes, etc.) beyond A(1,1). The $24 \times 200 - 1 = 4799$ is a compile-time computation which the programmer is unable to eliminate (without sacrificing the two-dimensional structure of the matrix A).

As we suggested earlier, compile-time computations generate constants which may in turn generate new compile-time computations. For example,

$$\begin{aligned} a &\longleftarrow 5 \\ b &\longleftarrow 7 \\ c &\longleftarrow a + b \\ &\ \vdots \\ d &\longleftarrow 3 \times c \end{aligned}$$

could be replaced with

$$\begin{aligned} a &\longleftarrow 5 \\ b &\longleftarrow 7 \\ c &\longleftarrow 12 \\ &\ \vdots \\ d &\longleftarrow 36 \end{aligned}$$

provided none of a, b, c is redefined in such a code sequence. The criteria for such *constant subsumption* or *constant propagation* are easily verified for any such expression occurring within a basic block.

Constants may be propagated between basic blocks by means of analysis analogous to that of the previous chapter, i.e., let $\alpha = x \circ y$ be an expression. If x and y are both equal to known constants at the point where α occurs in a basic block, α is a compile-time computation at that point. If either x or y is not defined as a constant in the block b and neither is defined by a run-time computation in the block b prior to the point at which α occurs, then α is a compile-time computation if and only if the arguments of α not defined in the block b are defined by the *same* constant on *every* path from the program entry block to the block b. This analysis would be quite extensive and, in general, would not be very worthwhile, however, due to its relatively low payoff in program efficiency.†

9.3. RUN-TIME OPTIMIZATIONS

Run-time computations are a different story. As we indicated in the previous chapter, it is possible to parse the program in such a manner that each binary expression $a \circ b$ is placed into a biunique correspondence with a

† The value-numbering scheme of Exercise 2 at the end of Chapter 8 is one method of implementing this analysis.

"variable" α. This identification immediately exposes certain redundancies within each block, viz., those expressions α_i which are defined at some point p_i in b and some subsequent point p_i' in b, while no computation relevant to α_i occurs between the two points. This is the case in the classic example of the solution of a quadratic equation. We have initially

$$x_1 \longleftarrow (-b + (b \uparrow 2 - 4 \times a \times c) \uparrow \tfrac{1}{2}) \div (2 \times a)$$
$$x_2 \longleftarrow (-b - (b \uparrow 2 - 4 \times a \times c) \uparrow \tfrac{1}{2}) \div (2 \times a)$$

This would be parsed as

$$\begin{aligned}
\alpha_1 &\longleftarrow b \uparrow 2\\
\alpha_2 &\longleftarrow 4 \times a\\
\alpha_3 &\longleftarrow \alpha_2 \times c\\
\alpha_4 &\longleftarrow \alpha_1 - \alpha_3\\
\Delta &\longleftarrow \alpha_4 \uparrow \tfrac{1}{2}\\
\alpha_6 &\longleftarrow -b + \Delta\\
\alpha_7 &\longleftarrow 2 \times a\\
x_1 &\longleftarrow \alpha_6 \div \alpha_7\\
\alpha_8 &\longleftarrow -b - \Delta\\
x_2 &\longleftarrow \alpha_8 \div \alpha_7
\end{aligned}$$

The obvious exploitation of the discriminant Δ and the denominator $2 \times a$ has been made, but we have only considered this code as occurring in the one block, b. In an example of this sort we need also to consider the environment in which such code is likely to occur. Most likely, the programmer would have been interested in the possibility of complex roots, so that his code may have resembled

if $b \uparrow 2 - 4 \times a \times c \geq 0$
  then $x_1 \longleftarrow (-b + (b \uparrow 2 - 4 \times a \times c) \uparrow \tfrac{1}{2}) \div (2 \times a)$
    $x_2 \longleftarrow (-b - (b \uparrow 2 - 4 \times q \times c) \uparrow \tfrac{1}{2}) \div (2 \times a)$
 else complex:root:routine
end

In this case the analysis from the preceding chapter would show that α_4 (viz., $b \uparrow 2 - 4 \times a \times c$) is available on entrance to the block b. The graph of this portion of the program, along with the identification of common subexpressions, becomes that shown in Fig. 9.1.

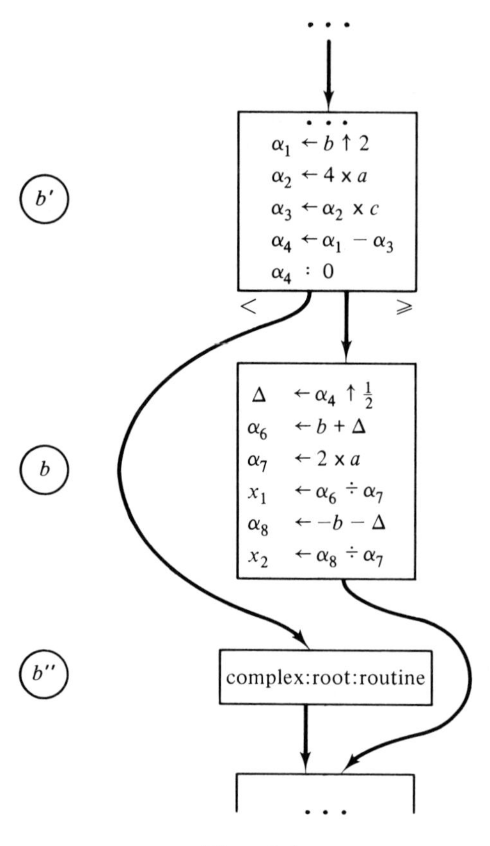

Figure 9.1

9.3.1. Code Motion

Now let us continue this example somewhat further. The code previously identified as complex:root:routine is probably of the form

$$\begin{aligned}
&Re\ x_1 \longleftarrow -b \div 2 \times a \\
&Im\ x_1 \longleftarrow |b \uparrow 2 - 4 \times a \times c| \uparrow \tfrac{1}{2} \div (2 \times a) \\
&Re\ x_2 \longleftarrow -b \div (2 \times a) \\
&Im\ x_2 \longleftarrow -|b \uparrow 2 - 4 \times a \times c| \uparrow \tfrac{1}{2} \div (2 \times a)
\end{aligned}$$

Analysis would show that α_4 was available on entrance to the block, which would therefore produce the code shown in Fig. 9.2. We note that α_7 is common to blocks b and b''. It follows that if α_7 were available on entrance to

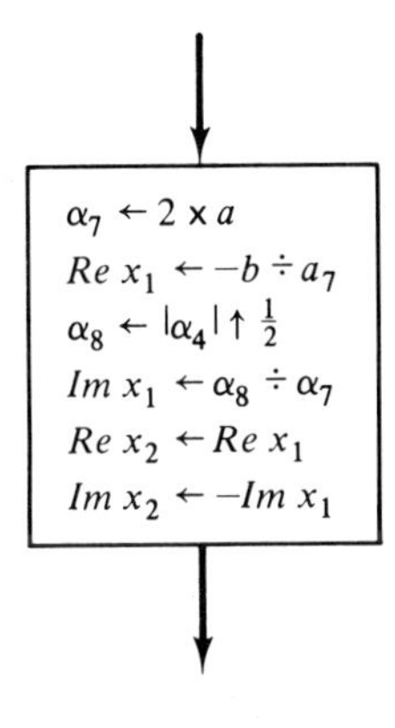

Figure 9.2

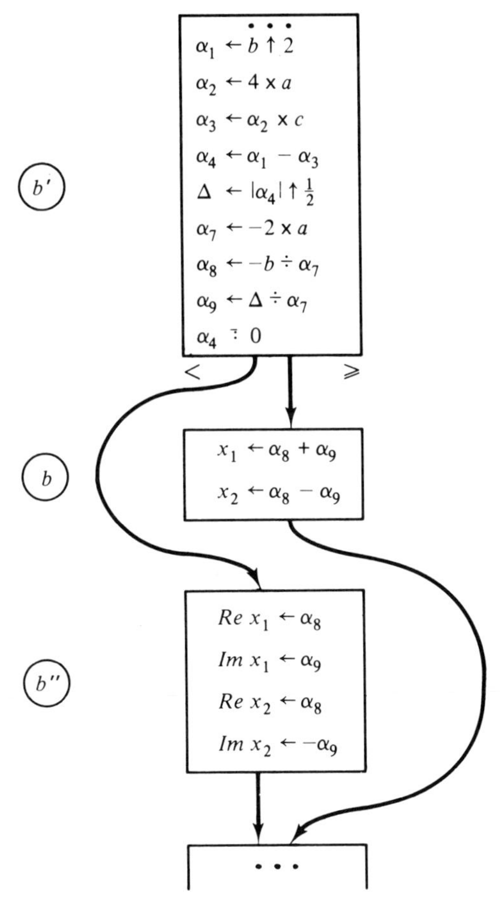

Figure 9.3

each of these blocks, the code defining α_7 could be deleted from them. α_7 could be *made* available on entrance to b and b'' by "*moving*" the definition of α_7 to block b', since b' is a back dominator of b and b''. Here, the optimization is one of space rather than one of time, since α_7 will be computed regardless of the sign of α_4. The latter form of optimization is referred to as *hoisting*.

Note that in the above example some rewriting of the principal expressions would have permitted greater compression of the code as the result of hoisting. Since the distributive law yields

$$\frac{-b \pm \sqrt{b^2 - 4ac}}{2a} = \frac{-b}{2a} \pm \frac{\sqrt{b^2 - 4ac}}{2a}$$

$-b \div \alpha_7$ could have been hoisted. Also, we could have expressed

$$\Delta \leftarrow |\alpha_4| \uparrow \tfrac{1}{2}$$

and branched according as $\alpha_4 \geq 0$ or $\alpha_4 < 0$. Doing so results in the code shown in Fig. 9.3, which consists of only 15 instructions.

9.4. BUSY VARIABLES

Let v be a programmer variable. It is apparent that if at some point in a program v is defined and v is not subsequently referenced, then the definition of v at that point is superfluous. Similarly, if v is defined at two points p_1, p_2 in the program and if there are no references to v on any path $\mu[p_1, p_2]$ and if all paths passing through p_1 necessarily pass through p_2, then the definition of v in p_1 is again superfluous. We note that such a definition may be of the form $v \leftarrow \alpha_k$, and α_k may be referenced (as a compiler variable) even though v is not used in any expressions prior to redefinition.

In such cases, the definition of v is called a *dead definition of* v and should be eliminated from the code. A *live definition of* v is a definition of v that is not dead, and which, therefore, cannot be eliminated from the program.

9.4.1. Busy Paths

We now proceed to develop a method for identifying dead definitions.

DEFINITION **9.1**

Let a programmer variable v be defined at a point p_1 in a program, and let $p_2 \in \hat{\Gamma} p_1$ be a point where v is used. If there does not exist a point p_3 ($\neq p_1$) in some *simple* path $\mu[p_1, p_2]$ in which v is redefined, we call μ a *v-busy path*. The variable v is said to be *busy* in any v-busy path. v is *not busy* immediately prior to a definition of v.

DEFINITION 9.2

Let $\mu = (b_1, \ldots, b_n)$ be a simple v-busy path which is not properly contained in any other v-busy path. The b_i are basic blocks. Then μ is a *maximal v-busy path.* The variable v is said to be *busy on exit* from any node $b_k \in \mu \sim \{b_n\}$ and *busy on entrance* to any node $b_j \in \mu \sim \{b_1\}$.

Note that there may exist several maximal v-busy paths which share common nodes. If $\mu = \mu[b_1, b_k]$ and $\nu = \nu[b_k, b_n]$ are two maximal v-busy paths, it follows that there is a definition of v in b_k which is preceded by a use of v in b_k. In such a case v is both busy on entrance and busy on exit from b_k, but $\lambda = \mu \cup \nu$ is not a v-busy path since there is at least one definition of v in b_k and hence at least one point in λ where v is not busy.

DEFINITION 9.3

A programmer variable v is *very busy on entrance* to a node b if either v is used in b prior to being defined in b or if v is not defined in b and every path leaving b contains a v-busy path head by b.

DEFINITION 9.4

A programmer variable v is *very busy on exit* from a node b if every path leaving b contains a v-busy path headed by b.

DEFINITION 9.5

A compiler variable α is *busy* (*very busy*) *on entrance* (*exit*) to (from) a node b if each of the arguments of α is busy (very busy) on entrance (exit) to (from) b and if α, considered as a programmer variable, is busy (very busy) on entrance (exit) to (from) b.

9.4.2. Algorithms for Finding Busy Paths

ALGORITHM 9.6

Let v be a programmer variable, and let $+$ and $\cdot$ be operators in a Boolean distributive lattice algebra. Then v is busy on entrance to a node b if and only if $E_b(v) = 1$, where

$$E_b(v) = \lim_{k \to \infty} E_b^{(k)}(v)$$

$$E_b^{(k+1)}(v) = E_b^{(k)}(v) + D_b(v) \cdot \sum_{c \in \Gamma b} E_c^{(k)}(v)$$

and $$E_b^{(0)}(v) = \begin{cases} 1 & \text{if } v \text{ is used in } b \text{ before being defined in } b \\ 0 & \text{otherwise} \end{cases}$$

and $$D_b(v) = \begin{cases} 0 & \text{if } v \text{ is defined in } b \\ 1 & \text{otherwise} \end{cases}$$

Finally, v is busy on exit from b if and only if $B_b(v) = 1$, where

$$B_b(v) = \sum_{c \in \Gamma b} E_c(v)$$

Proof. If v is referenced in b before being defined in b, v is busy on entrance to b by definition. In that case $E_b^{(0)}(v) = 1$, and since

$$E_b^{(k+1)}(v) \geq E_b^{(k)}(v) + D_b(v) \cdot \sum_{c \in \Gamma b} E_c^{(k)}(v),$$

we have $D_b(v) = 1$.

Assume that v is defined in b prior to being referenced in b. Then $E_b^{(0)}(v) = 0$ and $D_b(v) = 0$. Then v is not busy on entrance to b and

$$0 = D_b(v) = 0 + 0 \cdot \sum_{c \in \Gamma b} E_c^{(k)}(v).$$

Finally, assume that v is neither referenced nor defined in b, so that $E_b^{(0)}(v) = 0 = \tilde{D}_b(v)$. Then v is busy on entrance to b if and only if b lies on some v-busy path. If b lies on some v-busy path, then some immediate successor $c \in \Gamma b$ also lies on that v-busy path. Since every v-busy path is simple and therefore finite, there exists some final node $d \in \hat{\Gamma} b$ in which $E_d^{(0)}(v) = 1$. Hence there exists some node $d' \in \Gamma^{-1} d$ for which $E_d^{(1)}(v) = 1$, etc. Thus $E_b(v) = 1$ if and only if b lies on a v-busy path and v is busy on entrance to b.

Since v is busy on exit from b if and only if v is busy on entrance to some immediate successor of b, the result follows. ∎

ALGORITHM **9.7**

Let v be a programmer variable, and let $+$, $\cdot$, $E_b^{(0)}(v)$, and $D_b(v)$ be as in Algorithm 9.6. Then v is very busy on entrance to a node b if $V_b(v) = 1$, where

$$V_b(v) = \lim_{k \to \infty} V_b^{(k)}(v)$$
$$V_b^{(k+1)} = V_b^{(k)}(v) + D_b(v) \prod_{c \in \Gamma b} V_c^{(h)}(v)$$

and

$$V_b^{(0)}(v) = E_b^{(0)}(v) + D_b(v) \prod_{c \in \Gamma b} E_c^{(0)}(v)$$

Finally, v is very busy on exit from b if and only if $\hat{V}_b(v) = 1$, where

$$\hat{V}_b(v) = \prod_{c \in \Gamma b} V_c(b)$$

If $\Gamma b = \varnothing$, we define the empty product to be 0. (Why?)

Proof. $V_b^{(0)}(v)$ and $V_b^{(k)}(v)$ are simply elaborations of Definition 9.3 and are clearly monotonically increasing sequences. $V_b^{(k+1)}(v) > V_b^{(k)}(v)$ if and only if both $D_b(v) = 1$ (v is not defined in b) and $V_c^{(k)}(v) = 1$ for each immediate successor c of b (i.e., v is very busy on entrance to each of the immediate successors of b).

The limit exists due to the monotonicity of the sequence and the fact that once none of the $V_b^{(k)}(v)$ changes, none of the $V_b^{(k+1)}(v)$ can change.

Finally, $\hat{V}_b(v) = 1$ if and only if $V_c(v) = 1$ for each immediate successor c of b as a consequence of Definitions 9.3 and 9.4. ▮

9.5. BUSY VARIABLES AND HOISTING

While it is not much more useful to know that a programmer variable is very busy on exit from that block than just busy on exit from that block, the distinction for a compiler variable is significant. When a compiler variable α is very busy on exit from a block b, then we are guaranteed that α will be referenced before being killed regardless of whatever path the program takes from that block. If α is not available on exit from b (which is possible even if α is very busy on exit from b), then α must necessarily be computed at some point in the program, provided control passes through b. Hence, *if α were to be computed in b*, α would become available on entrance to every block on each α-busy path from b and would therefore become redundant in those blocks. The net result would be a saving in the space required for each such computation with no additional cost in execution time due to the hoisting we have achieved. In some cases it is even possible that a slight saving in execution time may result from this operation. The example involving the solutions to the quadratic equation illustrated hoisting to achieve conservation of space. In the examples below we shall illustrate hoisting and its effect on execution time.

Example 9.8

Consider the code shown in Fig. 9.4. Here we assume that $\alpha = q \circ r$ is not available on entrance to b_1 and is not killed in b_i, $i = 1, 2, 3, 4$. Clearly, α is very busy on exit from b_1. α is not available on entrance to b_4 due to the path (b_1, b_2, b_4) and must therefore be computed twice on the path (b_1, b_3, b_4). Since α is very busy on exit from b_1, it is clear that α will be computed sometime after exiting b_1 before α is killed. Therefore α may be hoisted to b_1, yielding the graph shown in Fig. 9.5. Now α is computed precisely once on either path, and a saving of one computation is achieved on the path (b_1, b_3, b_4).

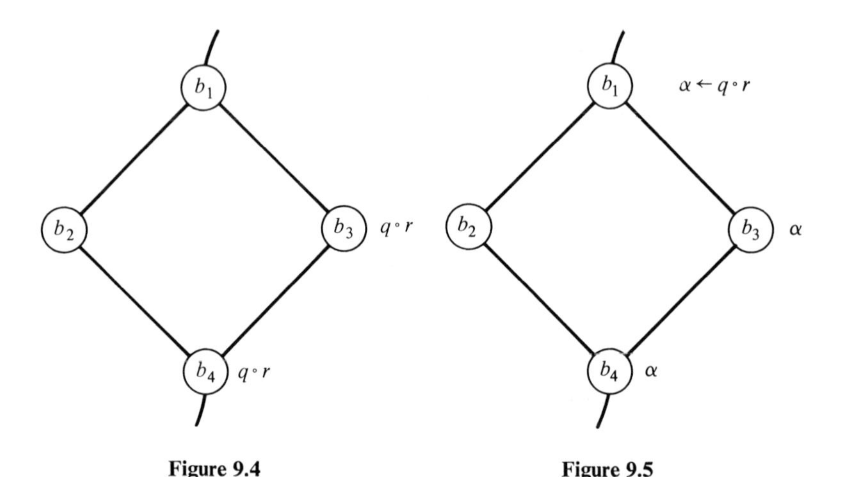

Figure 9.4

Figure 9.5

Example 9.9

Let $\alpha = q \circ r$, and consider Fig. 9.6. Here we again have that α is very busy on exit from b_1. However, if the program were transformed to the graph shown in Fig. 9.7, the net result would be a loss in space and time, for now $q \circ r$ must be stored into α in b_1 and then loaded and stored into x in b_4. In the code that would have been generated for Fig. 9.6, $q \circ r$ would not have been stored into α (since α is immediately killed in a subsequent statement) but directly into x.

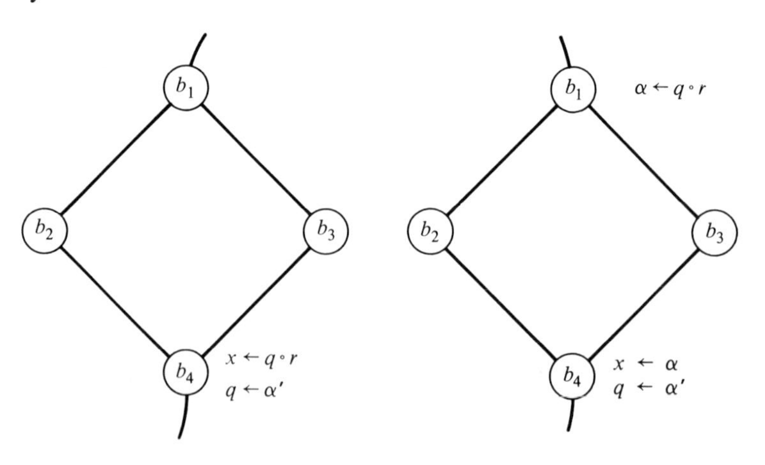

Figure 9.6

Figure 9.7

Example 9.10

Consider Fig. 9.8. Here α is very busy on exit from b_2, but hoisting α to b_2 does not generate any saving in code or time, since α is killed in b_5 and is

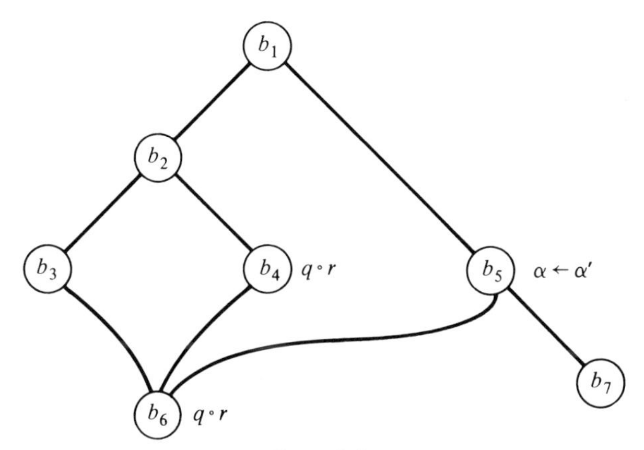

Figure 9.8

consequently not available on entrance to b_6 on the path (b_1, b_5, b_6). Were α to be hoisted to b_5 as well as b_2, an additional computation would result on the path $(b_1, b_5, b_7, \ldots)$ unless α were very busy on exit from b_5, etc. Thus it is apparent that while it is necessary that a variable be very busy on exit from a node in order that it be hoisted to that node, the condition is by no means sufficient to improve the code.

THEOREM **9.11**

Let $I = I(h)$ be an interval in the program graph, and let $\{y^M_{\alpha b} \mid b \in I(h)\}$ and $\{y^m_{\alpha b} \mid b \in I(h)\}$ be the initial solution set for the redundancy equations of Section 8.3.2, ranging over all expressions α. Then the expression α may be hoisted to the block $b \in I$ without loss of program efficiency only if

1. α is very busy on exit from b, but not available on exit from b.

2. Let $V = \{b' \in I \mid b' \text{ is on an } \alpha\text{-busy path from } b\}$, and for each $b' \in V$ let $G_{b'} = \{\Gamma^{-1}b'\} \sim V$. Then either $G_{b'} = \varnothing$ or

$$\prod_{g \in G_{b'}} Y^m_{\alpha g} = 1$$

3. α is referenced in at least two distinct nodes in V.

Proof. Since α is referenced in at least two distinct nodes $b_1, b_2 \in V$, each of which is back-dominated by b, it follows that placement of a computation of α at a point in b which would make $y_{\alpha b} = 1$ also makes $x_{\alpha b_1} = x_{\alpha b_2} = 1$ (since, by condition 2), α is available on exit from the sets G_{b_1}, G_{b_2}). Thus at least one computation of α is eliminated at no cost in program efficiency. The necessity of the conditions is obvious. ∎

For each interval of the program graph, the analysis of nodes from which an expression α is very busy on exit may be made, starting with the node with the highest-order number and working back toward the interval head. If a node b is formed to fit the criteria of Theorem 9.11, we define an associated vector P_b, where

$$P_{\alpha b} = \begin{cases} 1 & \text{if } (b, \alpha) \text{ satisfy Theorem 9.11} \\ 0 & \text{otherwise} \end{cases}$$

and redefine $y^M_{\alpha b} = y^m_{\alpha b} = 1$.

The solution of the redundancy equations then continues with the derived graphs as described in Section 8.5.2, except that following each step in the analysis, Theorem 9.11 is reapplied to the nodes of the derived graph and new vectors $P_{\alpha b}$ are introduced. In this manner, expressions are hoisted back as far as possible. Indeed, following this analysis, if $P_{\alpha b} = 1$ and $x_{\alpha b} = 0 = c_{\alpha b}$, then code for the computation of α must be added to b at an appropriate point. If, on the other hand, $P_{\alpha b} = 1$ and $x_{\alpha b} = 1$, then α has been hoisted above b, and no new code need be generated unless $k_{\alpha b} = 0$.

There remains the matter of determining the point within the block b where the code for α is to be added. If $k_{\alpha b} = 1$, then α may be computed at any point in b preceding a branch out of b, while if $k_{\alpha b} = 0$, the computation of α must be placed after the final computation relevant to α that is prior to the branches out of b. If expressions $\alpha_{i_1}, \alpha_{i_2}, \ldots, \alpha_{i_k}$ are being inserted into b and there is a dependency of α_{i_2} on α_{i_1}, then α_{i_2} must be inserted following α_{i_1} at points in b where the value of α_{i_1} is available to α_{i_2}.

9.5.1. Sinking Expressions

Example 9.12

Consider the graph in Fig. 9.9. Here we have an expression, $x \circ y$, which occurs on every path from b_1, and a node b_5 which is a forward dominator of each node in which $x \circ y$ occurs. We note that $x \circ y$ is *not* very busy on exit from b_1, even though (in this case) it is busy on exit from b_1. Consequently, $x \circ y$ cannot be hoisted to b_1.

However, since there are no definitions relevant to $x \circ y$ occurring between each of its occurrences and there are no computations following these occurrences in b_2, b_3, and b_4 which are dependant on $x \circ y$, then we move the computation to an appropriate point in b_5, where it can be computed once. In this manner a saving of space is realized. This process is called *sinking* an expression.

Expressions which are to be sunk are located by searching each block from its bottom to its top. In that manner, an expression α_j which is dependent on expressions $\alpha_{i_1}, \ldots, \alpha_{i_k}$ may be sunk before $\alpha_{i_1}, \ldots, \alpha_{i_k}$ are examin-

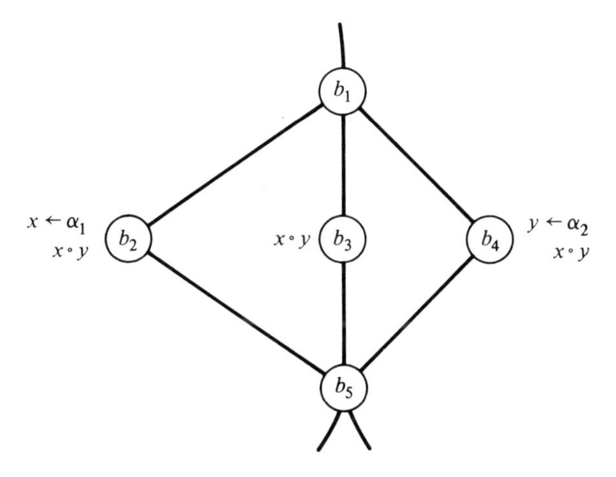

Figure 9.9

ed. Were the expressions examined in the opposite order, then α_{i_1} could not be sunk since α_j followed it and was dependent on it. We include among the computations dependent on α_j the expression $q \leftarrow \alpha_j$, where q is a programmer variable; $q \leftarrow \alpha_j$ may be sunk only if $q \leftarrow \alpha_j$ on each of the paths under consideration. Thus a larger number of sinkable expressions may be exposed in this manner.

We obtain the following analogue to Theorem 9.11.

THEOREM **9.13**

Let b be a basic block, and let b^* be the immediate back dominator of b. Let α be an expression which occurs on every *simple* path $\mu[b^*, b]$. Then α may be sunk into b without loss of program efficiency if and only if

1. α is not redundant on some path $\nu[b^*, b]$.

2. No computation relevant to α occurs between any candidate computation of α and b.

3. No computation dependent on α occurs between any candidate computation of α and b.

4. b is a forward dominator of each candidate computation of α.

The proof is left to the reader.

Historical note. Because of the transitory and artificial nature of the vectors P_b, they have been referred to by the author and his colleagues as "phoney baloney" vectors. For more obscure reasons, Theorem 9.11 is known as the Kosher Baloney Wine Sauce Theorem.

10 LOOP OPTIMIZATION: INVARIANT EXPRESSION REMOVAL AND REDUCTION IN STRENGTH OF OPERATORS

10.1. PROGRAM LOOPS

We shall refer to any strongly connected subgraph of the program graph as a *loop*. Loops are generally used to encode iterative processes and algorithms. A loop may be indicated by a key word in the higher level programming language or by a series of branches and tests that generate a closed path. Some loops may be implied by a higher-level language form rather than by open code; e.g., in K. E. Iverson's APL, the product of two matrices may be written as

$$C \longleftarrow A \overset{+}{\underset{\times}{}} B$$

to express $c^i_j \longleftarrow \sum_k a^k_i b^j_k$, which clearly iterates upon the three indices i, j, k.

Any two program loops are either disjoint, nested, or share a common intersection. Much of our attention will center on loops of the last two types, particularly on nested loops. Since the code in loops is generally executed several times every time the loop is entered, it is important that the code be as efficient as possible to reduce execution time. In some cases the modifications to the code that improve running efficiency are at variance with attempts at optimizing space utilization.

10.2. LOOP OPTIMIZATIONS

10.2.1. Unscrolling

Example 10.1

Consider the following loop:

$i \longleftarrow 1$

loop: $A[i] \longleftarrow k$

$i \longleftarrow i + 1$

if $i \leq 10$

then go to loop **end**

The index i must be incremented and tested 10 times in this loop. If the loop were "unrolled" or "unscrolled" as shown below, several savings would result:

$A[1] \longleftarrow k$

$A[2] \longleftarrow k$

$A[3] \longleftarrow k$

$A[4] \longleftarrow k$

$A[5] \longleftarrow k$

.
.
.

$A[10] \longleftarrow k$

Here, since the address of A is a constant, the addressing expressions $A[1]$, . . . , $A[10]$ are all compile-time expressions. No index variable is needed for the code; hence nothing need be incremented or tested. Finally, k would be loaded into an arithmetic register only once and then stored 10 consecutive times with no intervening code.

If space were a problem, even encoding the loop as shown below would produce some saving in execution time:

$i \longleftarrow 1$

loop: $A[i] \longleftarrow k$

$A[i + 1] \longleftarrow k$

$A[i + 2] \longleftarrow k$

$A[i + 3] \longleftarrow k$

$A[i + 4] \longleftarrow k$

$i \longleftarrow i + 5$

if $i \leq 6$

then go to loop **end**

Even better would be

```
        i ← 1
loop:   A[i] ← k
        A[i + 1] ← k
        A[i + 2] ← k
        A[i + 3] ← k
        A[i + 4] ← k
        if i = 1      (or i ≤ upper:bound)
              then i ← 6      (or i ← i + 5)
                    go to loop      end
```

In both these cases, the compiler can generate good code for subscripting of the form $A[i + n]$, where n is an integer. Again, most unrelated computations have been deleted. In the second case above, we moved the index increment to a place of lower execution frequency.

10.2.2. Code Motion in Loops

Moving code from a node with high execution frequency to a node with a relatively lower execution frequency is a worthwhile optimization.

Example 10.2

Consider the following code:

```
loop for i ← 1 to 1000 by 1:
      A[i] ← q∘r
      B[i] ← s*t      end
```

Here we note that the variables q, r, s, t are all constant throughout the body of the loop. Consequently, both $q \circ r$ and $s * t$ will be constant throughout the loop so that unnecessary computations will be made on 999 iterations of the code. The obvious enhancement (irrespective of unscrolling) is

```
c1 ← q ∘ r
c2 ← s * t
loop for i ← 1 to 1000 by 1:
      A[i] ← c1
      B[i] ← c2      end
```

10.2.3. Loop Fusion

Example 10.3

Consider the following code:

```
loop for i ←— 1 to 1000 by 1:
        A[i] ←— q ∘ r      end
        . . .
loop for j ←— 1 to 2000 by 1:
        B[j] ←— B[j] * t      end
```

Techniques employed in the previous examples could be used to *fuse* the two loops, producing

```
c1 ←— q ∘ r
loop for i ←— 1 to 1000 by 1:
        A[i] ←— c1
        B[2i − 1] ←— B[2i − 1] * t
        B[2i] ←— B[2i] * t      end
```

It will be shown below that the optimizing compiler would recognize the fact that i is incremented only in terms of itself and a constant (viz., 1) and that the multiplication of i by 2 can be achieved by initializing a variable $2i$ to the value 2 ($= 2 \times 1$), where 1 is the initial value of i, and simply adding 2 to $2i$ on each iteration of the loop.

10.2.4. Restrictions

Fusion of two loops may be accomplished only if certain requirements are satisfied, one of the most important of which being that neither loop may be exited prior to complete execution of the index range. If, for example, the second loop above had been

```
loop   for j ←— 1 to 2000 by 1:
       B[j] ←— B[j] * t
       if B[j] = 0
         then go to trap end      end
       . . .
trap:  . . .
```

and one of the $B[j]$ were equal to 0, then not all the $A[i]$ would be initialized to $q \circ r$.

Other obvious problems could result if there existed paths in the program graph which passed through one of the loops but not the other or if data in the second loop were dependent on computations made in the first loop or on the path from the first loop to the second loop.

Thus it is apparent that there exists a class of optimizations which are applicable only to loops. All these enhancements tend to reduce or eliminate certain computations which would have to be carried out on each iteration of the loop. Some of the transformations generate additional code outside of a loop as an overhead. If the loop is only executed once or twice, it may turn out that the cost of the overhead equals or even exceeds the cost of the original code, but if the loop is executed several times, especially hundreds of times, then the saving is significant. It will be useful for the user to bear these remarks in mind for the duration of this chapter.

10.3. LOOP CONSTANTS

DEFINITION **10.4**

A variable v is a *loop constant* (or *region constant*) *variable* for a loop L if v is not defined in any block in L.

DEFINITION **10.5**

An expression α is a *loop constant* (or *region constant*) *expression* for a loop L if none of its arguments is defined in any block in L.

PROPOSITION **10.6**

Let α be a loop constant expression for some loop L, and assume that α is not available on entrance to L. Let $F = \{f \in L \mid \tilde{L} \cap \Gamma^{-1}f \neq \varnothing\}$. Modify the program by introducing a new set of blocks $F' = \{f'_f \mid f \in F\}$, with

$$\Gamma^{-1}f'_f = \tilde{L} \cap \{\Gamma^{-1}f\}$$
$$\Gamma^{-1}f = \{f'_f\} \cup (L \cap \{\Gamma^{-1}f\})$$

Then a computation of α may be placed in each $f'_f \in F'$, and α will then be available on entrance to L.

Proof. F is the set of *entrance nodes* to L. The new set of nodes F' is introduced to provide a point outside of the loop for the computation of loop constants. Each node in F' is a back dominator of some entrance node to L and consequently control must pass through F' prior to executing any node in L. ∎

Normally $\#F = 1$, but the graph in Figure 6.5 possesses numerous multi-entry loops, including $\{e, h\}$ and $\{b, f, g\}$. In some but not all cases, the graph of a program possessing a loop with multiple entry points is irreducible. However, the structure shown in Fig. 10.1 is an interval containing a loop

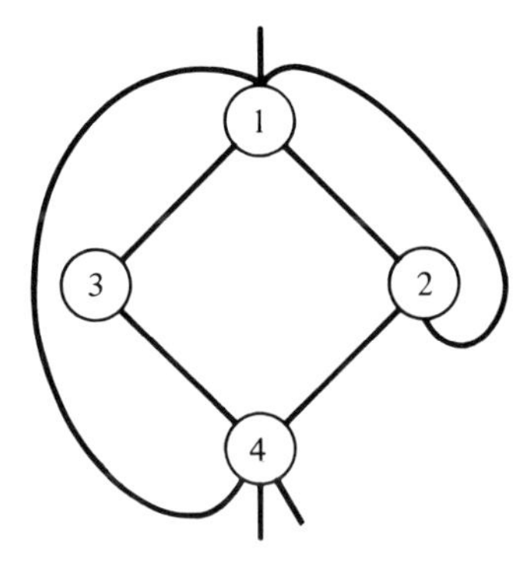

Figure 10.1

with multiple entry points, viz., $\{1, 3, 4\}$, which can be entered from either node $\{2\}$ or the predecessor's node $\{1\}$ external to the interval. An expression α constant in $\{1, 3, 4\}$ need not be constant in $\{1, 2\}$, so that α would have to be computed prior to entering $\{1, 3, 4\}$ from either $\{1\}$ or $\{4\}$.

10.4. PROPER LOOPS

To decrease the number of blocks that need to be introduced into the graph for the placement of loop constant expressions, we shall restrict the notion of *loop* with the following definition:

DEFINITION **10.7**

Let $I = I(h)$ be an interval in a graph G, and let S_c be the maximal strongly connected subgraph of I. A point $a \in S_c$ is an *articulation point* if $S_c \sim \{a\}$ is not strongly connected. A set $\{a_1, a_2, \ldots, a_k\} \in S_c$ is a *minimal articulation set* if $S_c \sim \{a_1, a_2, \ldots, a_k\}$ is not strongly connected, but for any $a_i \in \{a_1, a_2, \ldots, a_k\}$, $\{a_i\} \cup (S_c \sim \{a_1, a_2, \ldots, a_k\})$ is strongly connected.

EXERCISE

Prove or disprove: All articulation points are points of confluence but not conversely. Compare the definition following Proposition 3.11.

DEFINITION **10.8**

Let I and S_c be as above, and let I be ordered by Algorithm 5.3 (SNA). Let $L = \{l_1, l_2, \ldots, l_p\}$ be the back-latches of I, with $\mathcal{O}(l_1) < \mathcal{O}(l_2) < \cdots < \mathcal{O}(l_p)$. Then the *proper loops* of I are

1. S_c considered as a single loop.
2. Consider the cycles μ_i of S_c and let L_i be the set of back-latches of μ_i. The cycle μ_i is a proper loop if the elements of L_i possessing the highest-order numbers form a minimal articulation set.

Note that the cycles μ_i of Definition 10.8 are not necessarily prime cycles. Hence in Fig. 10.1 the only proper loop is $\{1, 2, 3, 4\}$, since the back-latch $\{2\}$ of the prime cycle $(1, 2)$ is not an articulation point.

10.4.1. Examples of Proper Loops

Example 10.9

The back-latches of the graph shown in Fig. 10.2 are $\{h, b, c, d, e, f\}$. The minimal articulation sets are $\{h\}$, $\{b\}$, $\{c\}$, $\{d, e\}$. Hence the proper cycles are

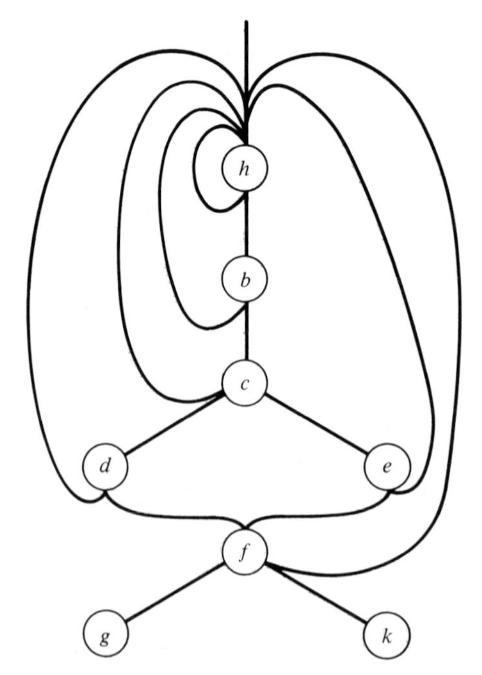

Figure 10.2

$$\mu_1 = \{h\}$$
$$\mu_2 = \{h, b\}$$
$$\mu_3 = \{h, b, c\}$$
$$\mu_4 = \{h, b, c, d, e\}$$
$$\mu_5 = \{h, b, c, d, e, f\}$$

Example 10.10

The graph shown in Fig. 10.3 is identical to that of Fig. 10.2 except for the arcs (b, e) and (b, d); $\{c\}$ is no longer an articulation point, since

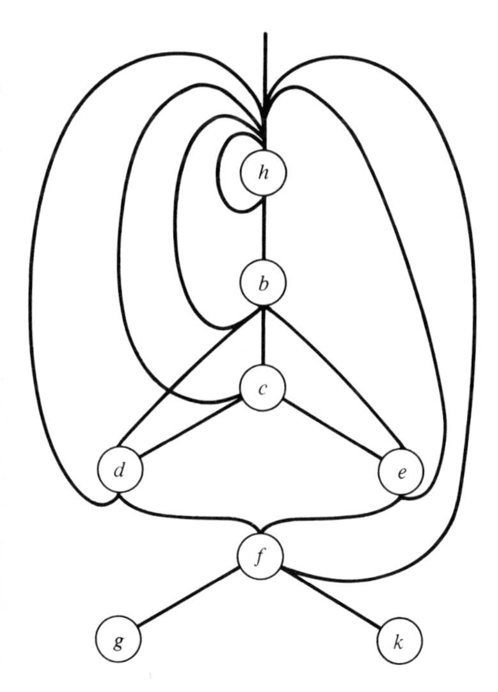

Figure 10.3

$\{h, b, d, e, f, g, k\}$ is connected. Indeed, the minimal articulation sets are now $\{h\}$, $\{b\}$, $\{d, e\}$ ($\{f\}$ does not affect strong connectivity). The proper loops are, therefore,

$$\nu_1 = \{h\}$$
$$\nu_2 = \{h, b\}$$
$$\nu_3 = \{h, b, c, d, e\}$$
$$\nu_4 = \{h, b, c, d, e, f\}$$

10.4.2. An Important Property of Proper Loops

Since the proper loops of an interval are nested within one another and since their back-latches form articulation sets, we are able to more compactly generate the set, F', of pseudoblocks described in Proposition 10.6.

PROPOSITION **10.11**

Let $I_p = I_p(h_p)$ be an interval containing proper loops

$$L_{p1} \subset L_{p2} \subset \cdots \subset L_{pk},$$

with back-latch sets $L_p^1, L_p^2, \ldots, L_p^k$, where

$$L_p^1 = \{l \in L_{p1} \mid h_p \in \Gamma l\}$$
$$L_p^i = \{l \in L_{pi} \sim L_{p,i-1} \mid h_p \in \Gamma l\}, \qquad i = 2, 3, \ldots, k$$

Then the node h_p can be embedded in the new set of nodes

$$F_p^* = \{f_{p0}, f_{p1}, \ldots, f_{pk}\},$$

where

$$f_{p0} = h_p, \qquad \Gamma f_{p0} \equiv \Gamma h_p$$
$$f_{pk} = \{\Gamma^{-1} h_p\} \sim I_p, \qquad f_{p,k-1} = \Gamma f_{pk}$$

and

$$f_{pi} = \Gamma L_p^{i+1} \sim (I_p \sim S_c), \qquad f_{p,i-1} = \Gamma f_{pi}, \qquad i = 1, 2, \ldots, k-1$$

Then $\{f_{pi}\}$ is precisely the set F' of Proposition 10.6.

Proof. Without loss of generality we may assume that G is a reducible graph, for if not, the methods of Chapter 6 may be applied. Hence all program loops are contained in some (derived) interval and may be combined with other program loops within the same interval to form proper loops.

Consider the innermost proper loop L_{p1} of interval I_p. Since the nodes of I_p are in the order of the SNA, it follows from Property 5.5 that the only entry point to L_{p1} is h_p. Identify the node f_{p0} with h_p and introduce the node f_{p1}, with $\Gamma^{-1} f_{p0} \equiv \{f_{p1}\} \cup L_p^1$. Then for control external to L_{p1} to enter L_{p1}, f_{p1} must be traversed. However, while inside L_{p1}, control is transferred back to $f_{p0} = h$. Hence, if a loop constant, α, for L_{p1} is placed inside f_{p1}, α will be redundant within L_{p1}.

Now assume that the result has been proved for all $L_{pi} \subset I_p$, $1 \leq i < t \leq k$, and let $f_{pt}, f_{p,t-1}, \ldots, f_{p0}$ be in accordance with the hypothesis. Then all proper loops $L_{pi} \subset L_{pt}$ $(i < t)$, and any expression β which is constant within L_{pt} is constant (although not necessarily referenced) within the L_{pi}. Since control external to L_{pt} is transfered to the loop only through f_{pt}, while the back-latches transfer control only to $f_{p,t-1}$, it follows that f_{pt} is the immediate back dominator of L_{pt}. Hence placing β in f_{pt} renders β redundant throughout L_{pt}. ∎

10.4.3. Expanded Intervals

DEFINITION **10.12**

An interval possessing proper loops that has been modified according to Proposition 10.11 is an *expanded interval.*

Example 10.13

The graph of Fig. 10.3 would be transformed to that shown in Fig. 10.4 to satisfy the requirements of Propositions 10.6 and 10.11. Here those expressions constant within $v_3 = \{h, b, c, d, e\}$ are computed in f_3 and hence become redundant within v_3. If some expression α is constant within v_4 as well as within v_3, its insertion in f_4 makes α redundant in f_3. Thus it is possible for one or more of the f_i to contain no code after some point in the analysis.

10.4.3.1. A Modification to the Redundancy Equations

The alert reader will note that Fig. 10.4 is not an interval. It is not desirable to repartition the graph after expanding the interval head, since the loop structure within the prior interval is, of necessity, single-entry and the information from the redundancy equations may be extended to the new nodes with minimal need for further analysis since expressions in blocks $f_1, \ldots, f_k$ are computed but not killed. This is achieved via

DEFINITION **10.14**

$$x_{\alpha f_{pk}} = \prod_{t \in \{\Gamma^{-1} h_p\} \sim I_p} y_{\alpha t}$$

$$y_{\alpha f_{pk}} = x_{\alpha f_{pk}} + c_{\alpha f_{pk}}$$

Similarly,

$$y_{\alpha f_{p,i-1}} = c_{\alpha f_{p,i-1}} + y_{\alpha f_{pi}} \prod_{l \in L^i_p} y_{\alpha l}, \qquad i = 1, 2, \ldots, k-1$$

PROPOSITION **10.15**

Let I be an expanded interval possessing k proper loops $L_1, L_2, \ldots, L_k$ with associated back-latch sets $L^1, L^2, \ldots, L^k$.

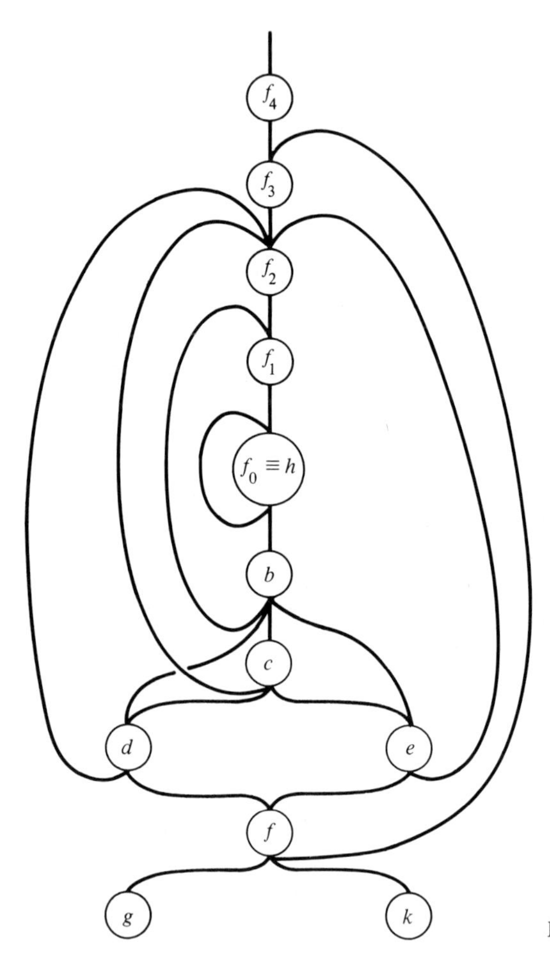

Figure 10.4

Let $y_{\alpha f_j}$, $j = 0, 1, \ldots, k$ be defined for all expressions α according to Definition 10.14. Let L_t be the proper loop with the least index in which the expression β is constant, and assume that β is not redundant within L_t. Then if a computation of β is placed in f_t and we define $c_{\beta t} = 1$, and for $i = 0, 1, \ldots, t$ set $y_{\beta i} = 1$, then we will have that β is redundant within L_t, and for every node $m \in L_t \sim \{f_0, f_1, \ldots, f_t\}$, the new solution of the redundancy equations is $x_{\beta m} = y_{\beta m} = 1$.

Proof. It is clear that since f_t back dominates the nodes of L_t and β is constant within L_t, placement of a computation of β within f_t makes β redundant within all of L_t. Since β is a loop constant, $k_{\beta m} = 1$ for all $m \in L_t$. The SNA guarantees that all of the predecessors of any node $m \in L_t$ will be processed prior to m itself and, therefore, $x_{\beta m} = \prod_{d \in \Gamma^{-1} m} y_{\beta d} = 1$ by definition. ∎

Proposition 10.15 permits routine redefinition of the $y_{\beta m}$ to 1 for all loop constants β in all nodes $m \in L_t$. Hence, the only redundancy variables in I requiring recomputation are those for the nodes $n \in I$ whose order indices exceed those of the nodes of L_t. These recomputations will yield new values for the availability on exit of certain expressions from the interval I which can be taken into account when the processing proceeds to the derived intervals of the program. This will be described presently.

10.5. OPTIMIZATION BY CODE MOTION

10.5.1. Motion of Loop Constants

The systematic identification of loop constants starts with the innermost proper loop (L_1) of each interval. Those loop constants which are not available on entrance to L_1 are placed in f_1 and the modifications of Proposition 10.15 are made. Next, L_2 is examined for loop constants. Those which are not available on entrance to L_2 (and hence f_2) are placed in f_2, etc., through all the proper loops of I working outward to L_k.

If, after this procedure has terminated, it is ascertained that a loop constant β is computed downward in blocks $f_i, f_{i+1}, \ldots, f_{i+r}$, the computation of β can be removed from $f_i, f_{i+1}, \ldots, f_{i+r-1}$ since β is available on entrance to each of these blocks from f_{i+r}, which back-dominates them. Thus only one computation need be made of each loop constant for all of I.

Example 10.16

Consider the code sequence in Fig. 10.5, from which it is clear that $a \circ b$ is not available on entrance to the interval headed by h; nor is $a \circ b$ a loop constant, since it is killed in b_3 before being recomputed. Finally, if the back-latch is taken, $a \circ b$ must be computed unnecessarily in h since it is available

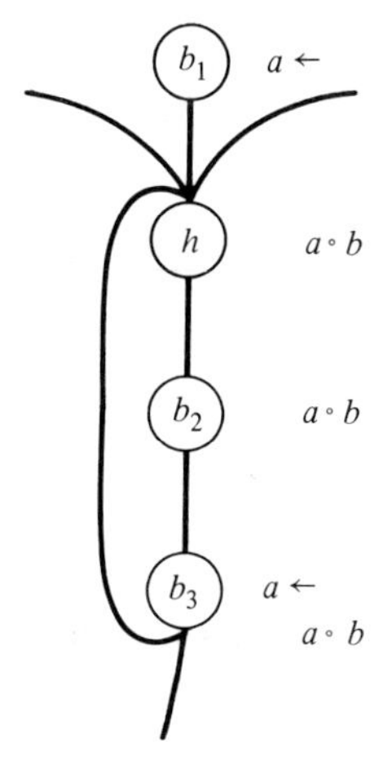

Figure 10.5

on exit from b_3. A compromise solution is exhibited in Fig. 10.6. Here the compiler variable α has been used to store a fresh computation of $a \circ b$ in the expanded interval head f_1 so that $a \circ b$ is available on entrance to the loop. Since α is also available on exit from b_3, we have successfully reduced the number of unnecessary computations of a loop variable.

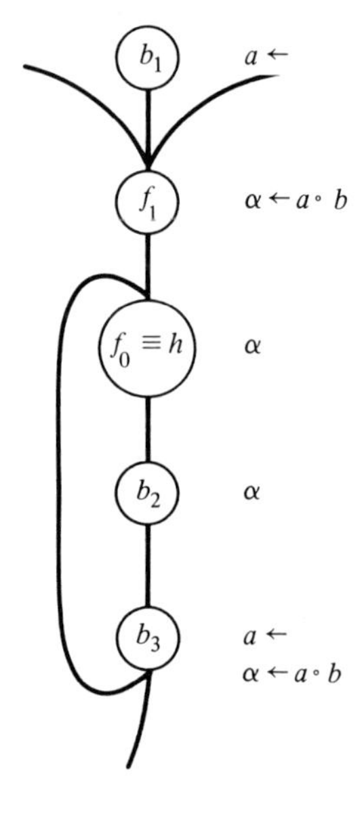

Figure 10.6

10.5.2. Motion of Non Loop Constants

PROPOSITION **10.17**

Let I be an expanded interval containing a proper loop L_t, and let α be a compiler variable which is not available on entrance to f_{t-1}, which is not a loop constant for L_t, and is available on exit from the back-latches L^t. Let α be busy on exit from f_{t-1}. Then α may be computed in f_t, thereby becoming redundant in those blocks of L_t lying on α-busy paths from f_{t-1}.

Proof. Since α is available on exit from the nodes of L^t by hypothesis and is available on exit from f_t by construction, α is available on entrance and, therefore, on exit from f_{t-1}. Since α is busy on exit from f_{t-1}, there exists a definition-free path from f_{t-1} to a use of α. ∎

10.5.3. Optimization Without Adverse Side Effects

THEOREM **10.18**

Let α be a compiler variable which satisfies the hypothesis of either Proposition 10.15 or 10.17 for the proper loop L_t. Then computing α in f_t

produces an optimization which has no adverse side effects on the program if α is very busy on exit from h, provided L_t is iterated more than once.

Proof. (by counterexamples). Assume not. Then the program may be slowed down by such a move, for consider the proper loop shown in Fig. 10.7.

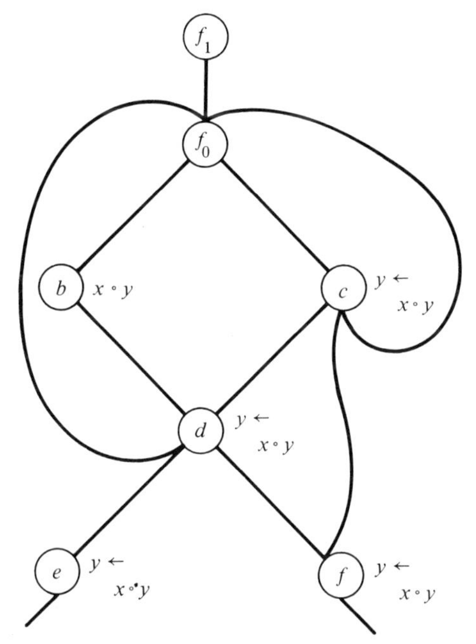

Figure 10.7

Here $x \circ y$ is busy on exit from f_1 and available on exit from the back-latches $\{c, d\}$. However, if the path $[f_0, b, d]$ is *never* taken, the computation is unnecessary and time-consuming.

Furthermore, the transformation *may* produce unwelcome side effects. For example, let $\circ$ be the division operator. It is well known that $x \div y$ will produce a division interruption if $|y| < \epsilon$ for a sufficiently small number ϵ. The branch in node f_0 may well be a test to prevent just such an operation from occurring. (If we, in our zeal to improve the efficiency of a program, proceed with an imagined carte blanche, we may well find our name scorned and our compilers unused!)

If, however, α is very busy on exit from f_0, then it follows that α will be computed in L_t before it is defined, regardless of the actual flow of control in the loop. Since α will be computed at least once in the loop, an optimization has been achieved unless the loop is not iterated, in which case the loss is an unnecessary store and load of α. (A more comprehensive treatment of this topic may be found in Chapter 11.) ∎

COROLLARY **10.19**

The conditions of Theorem 10.18 are satisfied if α is busy on exit from L and an articulation node (or a point of confluence) of L_t lies on an α-busy path from h.

10.6. REDUCTION IN STRENGTH OF OPERATORS

DEFINITION **10.20**

Let L_t be a proper loop contained in an interval I, and let ρ be an arbitrary variable defined within L_t. Then ρ is an *inductive variable* (or *recursive variable*) if and only if every definition of ρ within L_t is of the form

$$\rho \longleftarrow \pm \rho \pm \Delta_i,$$

where the Δ_i are all loop constants for L_t.

DEFINITION **10.21**

Let $\circ$ and $*$ be two operators which are so related that $*$ is defined in terms of (possibly repeated applications of) $\circ$. Then representing an operation defined in terms of $*$ exclusively in terms of $\circ$ is a *reduction in strength of* $*$.

Operators of the category referred to in Definition 10.21 include integral multiplication (which is a sequence of repeated additions) and integral exponentiation (which is a sequence of repeated multiplications).

10.6.1. One Inductive Variable

We shall next give several examples of the reduction in strength of various operators. The reader should be capable of generalizing from these examples.

Example 10.22

It is clear from Fig. 10.8(a) that i is an inductive variable. We note that on successive iterations of the loop the values of $10 \times i$ are

$$\begin{aligned}
&10 \times 1 \\
&10 \times 3 = 10 \times (1 + 2) = 10 \times 1 + 10 \times 2 \\
&10 \times 5 = 10 \times (3 + 2) = 10 \times 3 + 10 \times 2 \\
&\qquad\quad = 10 \times 1 + 10 \times 2 + 10 \times 2 \\
&10 \times 7 = 10 \times 1 + 10 \times 2 + 10 \times 2 + 10 \times 2
\end{aligned}$$

etc.

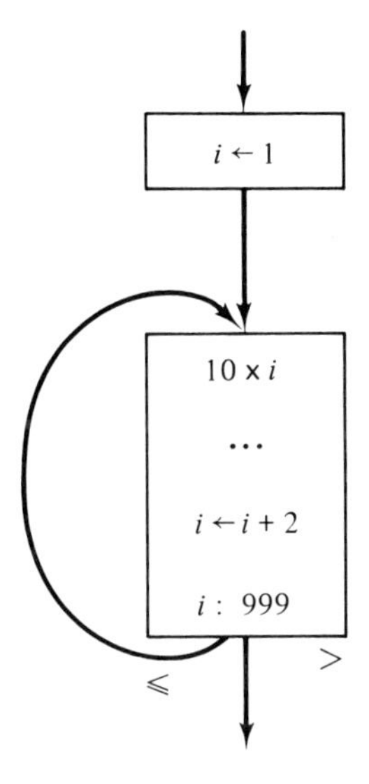

Figure 10.8(a)

Thus the code in Fig. 10.8(b) produces the equivalent result, and since addition is faster than multiplication, an increase in program speed is achieved.

It may happen that the only uses of i are the increment by 2 and the comparison against 999 at the bottom of the loop. If that is the case, i is not really necessary to the loop, and the code shown in Fig. 10.8(c) might well be preferred.

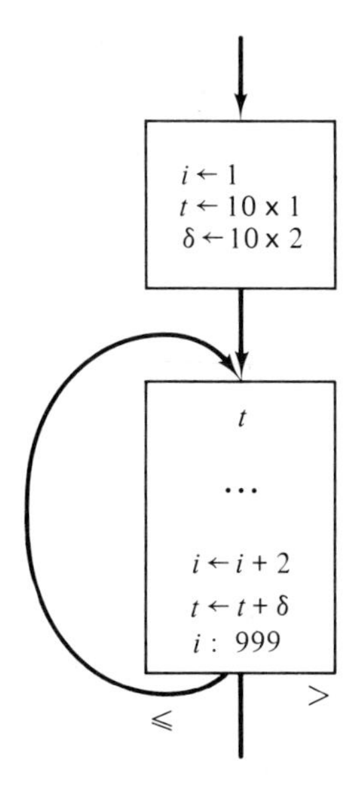

Figure 10.8(b)

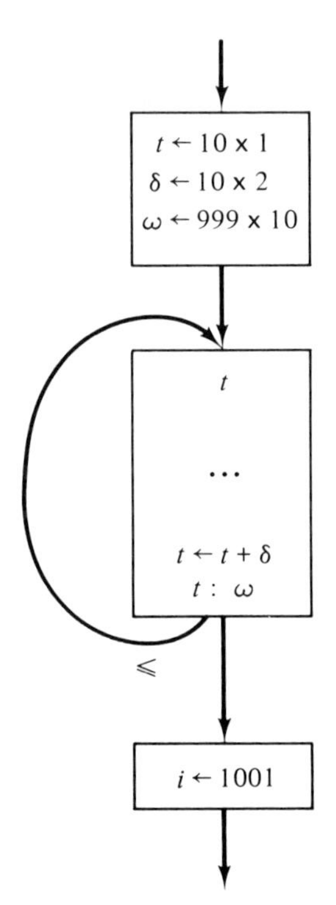

Figure 10.8(c)

Finally, if i is not busy on exit from the loop, the definition of i in the successor block is dead and should be deleted.

Example 10.23

The procedure employed in Example 10.22 is sufficiently general to handle a proper loop in which the recursive variable is defined several times and is multiplied by numerous loop constants. Consider the code in Fig. 10.9. It is assumed that k_1, k_2, c_1, c_2, c_3 are all loop constants. It is clear that on the first pass through the loop $k_1 \times i = k_1 \times i_0$ and $k_2 \times i = k_2 \times i_0 + k_2 \times c_1 = k_2 \times (i_0 + c_1)$ or $k_2 \times i = k_2 \times i_0 + k_2 \times c_2$. On the next iteration, $k_1 \times i = k_1 \times (i_0 + c_1 + c_3) = k_1 \times i_0 + k_1 \times c_1 + k_2 \times c_3$ or $k_1 \times i = k_1 \times (i_0 + c_2 + c_3) = k_1 \times i_0 + k_1 \times c_2 + k_1 \times c_3$, while $k_2 \times i = k_2 \times (i_0 + c_1 + c_3 + c_1) = k_2 \times i_0 + k_2 \times c_1 + k_2 \times c_3 + k_2 \times c_3$ or $k_2 \times i = k_2 \times (i_0 + c_1 + c_3 + c_2)$ or $k_2 \times i = k_2 \times (i_0 + c_2 + c_3 + c_1)$ or $k_2 \times i = k_2 \times (i_0 + c_2 + c_3 + c_2)$. The pattern is clear, and we obtain the code shown in Fig. 10.10.

Again, i may no longer serve a purpose in the loop, so that the expressions involving i might be deleted. The multiplications have all been removed from the loop, at a cost of numerous additions and several additional instructions. On each iteration of the modified loop two additions are required in place of

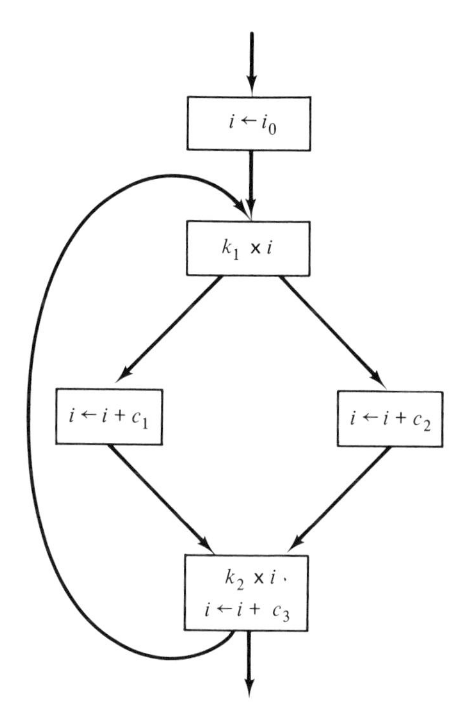

Figure 10.9

each multiplication. Thus the value of this modification is dependent on the time required on the particular computer for the various operations. For certain computers, the original code may be preferable. It should be noted that the initialization code is only executed once and that in many of the indicated operations there may be compile-time computations.

We note in passing that had the code involved exponentiation of the form k_1^i, then for $i \leftarrow i_0 + \Delta$, $t \leftarrow k_1^i = k_1^{(i_0+\Delta)} = k_1^{i_0} \times k_1^{\Delta} = t \times \delta$, where $k_1^{i_0}$ and $\delta = k_1^{\Delta}$ are loop constants. Hence precisely the same procedure would be applicable for the reduction in strength of integral exponentiation as is employed in the reduction in strength of integral multiplication. (We note that our restriction to integral multiplication and exponentiation is one of conservatism based on problems associated with certain computers. Given variables whose values fall in the proper range, the algebraic properties of the real numbers should not be adversely affected by problems of precision.)

10.6.2. Two Inductive Variables

Example 10.24

Figure 10.11(a) shows the reduction in strength of two recursive variables. On the nth iteration, we have $i \times j = (i_0 + mc_1)(j_0 + (n - m)c_2) = i_0 j_0$

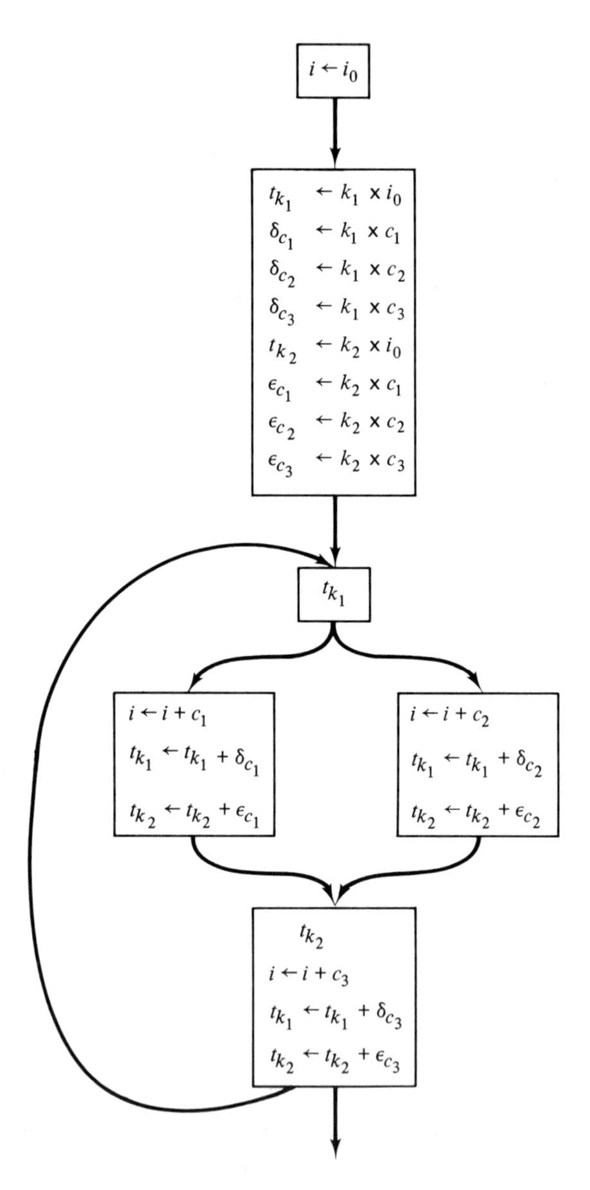

Figure 10.10

$+ mc_1 j_0 + (n - m)c_2 i_0 + m(n - m)c_1 c_2$. Calling this quantity t, we have that on the $(n + 1)$st iteration either $(i_0 + (m + 1)c_1)(j_0 + (n - m))(c_2) = t + c_1 j_0 + (n - m)c_1 c_2$ or $(i_0 + mc_1)(j_0 + (n - m + 1)c_2) = t + c_2 i_0 + mc_1 c_2$. Finally, we note that on the nth iteration, if either $m = 0$ or $m = n$, we have that $i \times j$ is, respectively, of the form $i_0 j_0 + nc_2$ or $i_0 j_0 + nc_1$. Com-

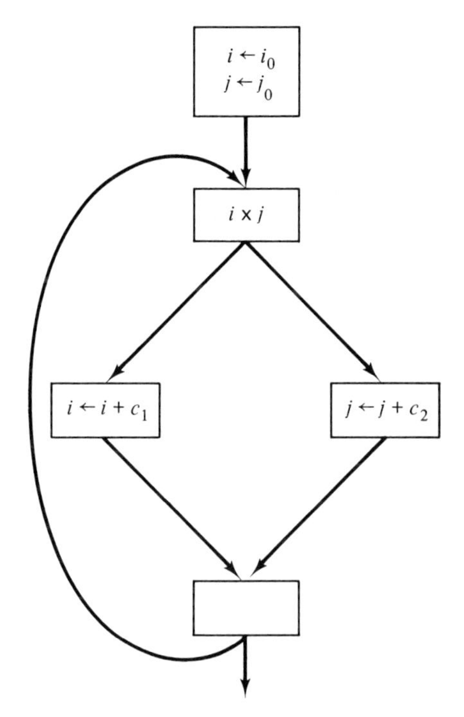

Figure 10.11(a)

bining these observations, we obtain the configuration shown in Fig. 10.11(b).

Had the multiplication been of the form $k \times i \times j$, an appropriate parsing would have reduced the operation to the form $k \times t$, and t is clearly a recursive variable. If k were a loop constant, then $k \times t$ could be reduced to a series of additions as depicted in Fig. 10.11(c).

However, the recursive variable t is no longer used in the loop. If i and j are also no longer of use, the code may be further modified to the form presented in Fig. 10.11(d). The code probably consists of a large number of compile-time computations, so that even less computation occurs at execution time than meets the eye.

Example 10.25

The square of a recursive variable is of the form $i^2 = (i_0 + \Delta)^2 = i_0^2 + 2i \times \Delta + \Delta^2$. Hence Fig. 10.12(a) reduces to Figure 10.12(b).

The majority of expressions actually occurring in proper loops which are subject to reduction in strength result from subscript computations generated by the compiler itself. Since the code occurring in loops normally includes numerous references to the same subscript expression, reduction in strength can often result in greatly improved compiled code.

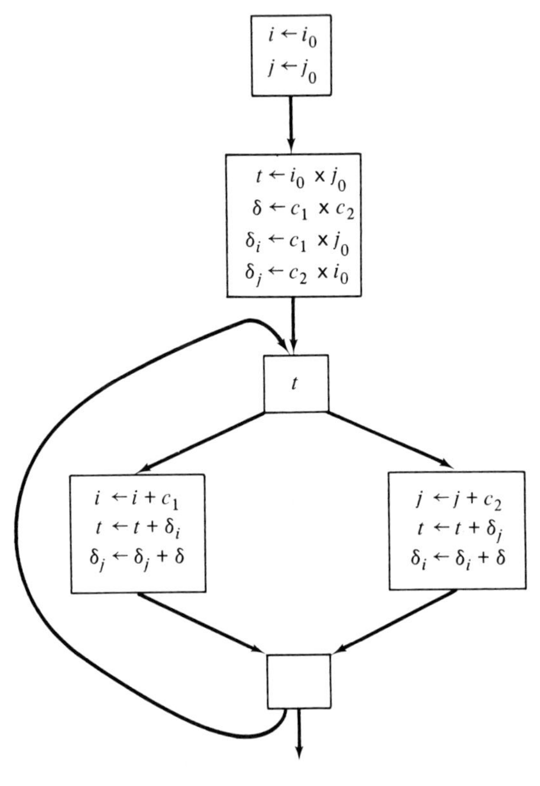

Figure 10.11(b)

10.6.3. Elimination of Inductive Variables

As was illustrated in Example 10.22, it may happen that after reduction in strength optimizations are carried out, the only remaining uses of the recursive variables are involved in the test at the bottom of the loop where the decision is made to iterate again or not. If the test is made against a loop constant, then the recursive variable can be eliminated and the test revised in terms of the newly generated recursive variable.

10.7. SUMMARY

In summary, the proper loops of each interval are identified and, in turn, optimizations are performed first on the most nested proper loop, L_1; then on that part of the smallest proper loop, L_2, containing L_1 (to wit $L_2 \sim L_1$); and so forth. [This is proper, since a loop constant for L_2 is a loop constant for L_1 but not conversely. Hence, if α were a loop constant for L_1, α would be

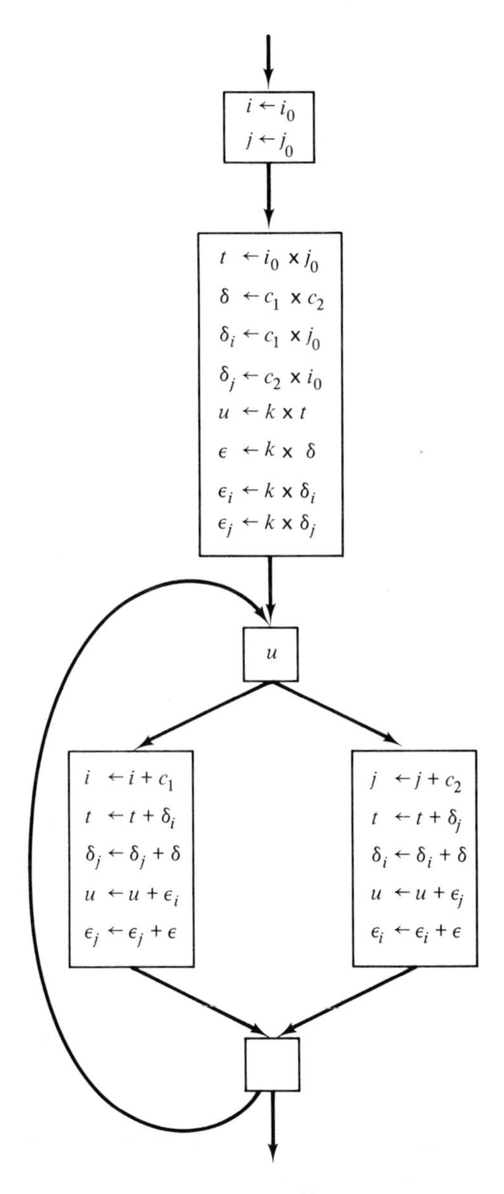

Figure 10.11(c)

placed in f_1. If α were also a loop constant for L_2, α could be removed from $L_2 \sim L_1$ (including f_1) and placed in f_2, etc.] When each of the intervals has been processed, the proper loops of the intervals in the derived graph are processed in precisely the same manner and so forth through all the derived graphs of the program.

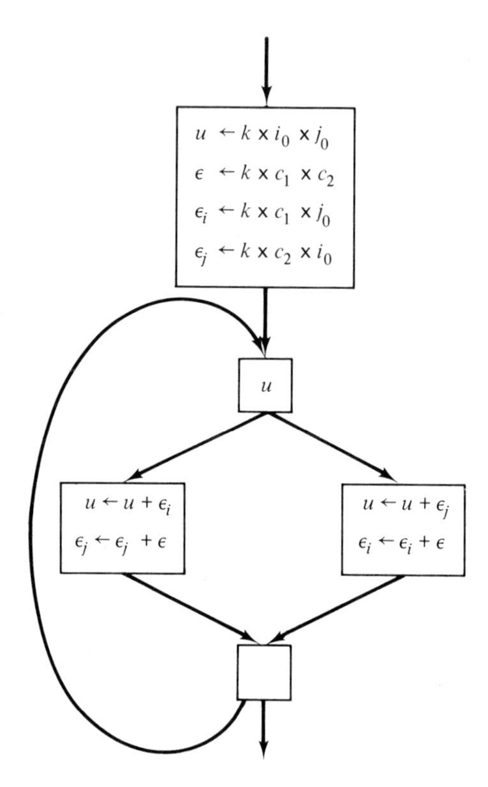

Figure 10.11(d)

The optimizations of this chapter are, of course, to be combined with those of the other chapters for maximum effect. In fact, reduction in strength often makes it necessary to eliminate new redundant subexpressions, dead variables, etc. The dependencies of certain optimizations on other optimizations will be discussed in Appendices II and III.

EXERCISES

1. Produce an algorithm for recognizing the inductive variables in a proper loop.

2. Give a necessary and sufficient condition that an inductive variable be removed from a loop following a reduction in strength optimization.

3. On some computers, iterative loops run more efficiently if the inductive variable starts at a maximum value and is consecutively decremented until it reaches 0. Given an iterative loop in which the inductive variable is incremented, what are the necessary and sufficient conditions that permit transforming the loop so that it runs "backwards?"

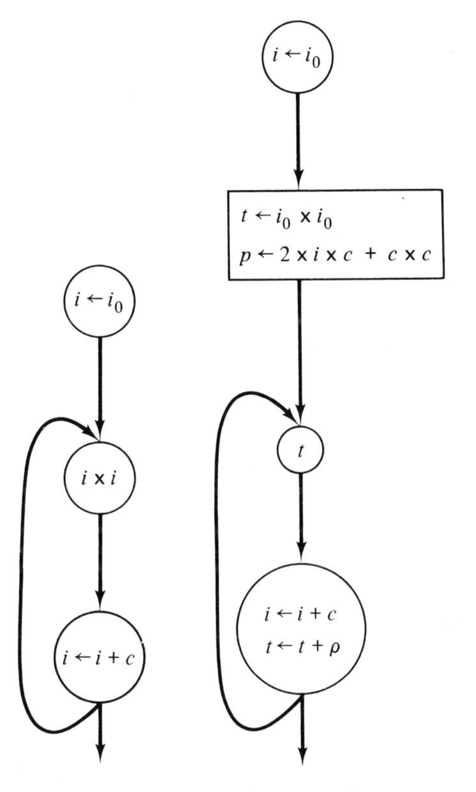

Figure 10.12(a) **Figure 10.12(b)**

11 SAFETY, PROFITABILITY, AND EXECUTION FREQUENCY CONSIDERATIONS

11.1. PROGRAM EQUIVALENCE

In previous chapters we have touched lightly on problems of computational equivalence between optimized and unoptimized code. We have also mentioned the possibility that an optimization transformation such as hoisting might introduce a computation into the control flow which would not otherwise have occurred, thus slowing the program somewhat. It is clear that there are certain kinds of equivalence which should be maintained between the original program and the optimized program: Both programs should achieve the same numerical results, the optimized program should run no slower than the unoptimized program, the optimized program should not produce machine interruptions that the original program did not produce, etc.

In this chapter we shall discuss these problems at greater length. The discussion cannot produce a concise mathematical theory, due in part to the recursive unsolvability of the problem of computational equivalence and in part to certain problems which occur only on specific computers.

11.2. SAFETY

We turn first to the problem of safety, i.e., guaranteeing a form of computational equivalence between the original program and the optimized version of the program. We start with

DEFINITION **11.1**

Let α be an expression computed at a point p in a program whose control flow graph is $G = (X, \Gamma)$. Then it is *strictly safe* to place a computation of

α at a point p' if and only if p' is a back dominator of p from which α is very busy on exit, and if α were available on exit from p', then α would be available on entrance to p.

If it is strictly safe to move a computation of α from p to p', it does not necessarily follow that a computation of α at p' will not cause a machine interruption but rather that any interruption that would occur at p' would necessarily occur at p if α were not moved to p'. Thus the criteria for strict safety are identical to those for hoisting, which are given in Theorem 9.11.

11.2.1. Precision

We are interested in performing transformations to code rather than limiting ourselves to those which are strictly safe. No computation introduced by the optimizing compiler should be capable of causing an interruption, and we must, therefore, be conservative in all optimization transformations. For example, the algebraic identity $\sum_{i=1}^{n} ca_i \equiv c \sum_{i=1}^{n} a_i$ does not necessarily hold for machine computations due to the possibility of introducing a computer-model-dependent error such as overflow, underflow, loss of precision, etc. Yet it is clear that an optimization would result from replacing a term-by-term constant multiplication with one multiplication of the sum. For lack of outside information, the optimizing compiler should leave the code alone, but as will be seen, information about the frequency of execution of the basic block containing this piece of code may influence us to produce potentially unsafe code in this instance.

11.2.2. Interruptions

Similarly, there is a temptation to improve the novice's code for the computation of an arithmetic average from $\sum_{i=1}^{n} (a_i \div n)$ to $(\sum_{i=1}^{n} a_i) \div n$ (which could occur accidentally in APL\360 where one could write $+/A \div N$ instead of the more efficient $(+/A) \div N$). Since the division by n would occur in either case, it is clear that there is no danger of our introducing a zero-divide interruption to the code. The remaining question is whether the programmer deliberately wrote the code in the inefficient form for reasons of precision, for if so, we again have no justification for making the transformation.

Recall that a computation, α, that occurs in a block, b, which is a point of confluence in some loop, may be moved to a back dominator $d \leq b$ provided that every simple path $\mu = \mu[d, b]$ is an α-busy path and that for every simple cycle $\nu = \nu[b, b]$ either $\mu \subset \nu$ or α is computed downward in each block $p \in \nu$ in which α is killed. Such a transformation is obviously strictly safe.

If α is some operation whose execution cannot cause a program interruption, α can be computed in any back dominator which satisfies the above criteria. There may, however, be a degradation in computational efficiency if the block in which α was originally located is not a point of confluence, since the code motion may force a computation to be performed which would not otherwise occur.

It is critical that α be incapable of causing a machine interruption when it is moved to the back-dominating block, as the following example illustrates.

Example 11.2

Let β, γ, and φ be functions and consider the code

```
loop for i ← 1 to l:
    loop for j ← 1 to m:
        if β(i) ≥ γ(i)
            then a_ij ← φ(β(i))
                ...
            end
        end
    end
```

Note that β, γ, and φ are independent of the loop variable j and are consequently constant within the inner loop. However, $\varphi(\beta(i))$ is not evaluated in a point of confluence of the inner loop, and so it is not strictly safe to move the computation to a point outside the loop. Thus the propriety of moving the computation $\varphi(\beta(i))$ is dependant on the nature of the function φ.

Consider the following *interpreted* code, where β, γ, and φ are defined functions:

```
loop for i ← 1 to l:                    (11.2.1)
    loop for j ← 1 to m:                (11.2.2)
        if b_i ≥ 0                      (11.2.3)
            then a_ij ← √b_i            (11.2.4)
                ...                     (11.2.5)
            end                         (11.2.6)
        end                             (11.2.7)
    end                                 (11.2.8)
```

If the code $t \leftarrow \sqrt{b_i}$ were inserted following line (11.2.1), then a program interruption would occur in those cases where $b_i < 0$. The only safe way such code could be optimized would involve removing the test on b_i from the loop along with the computation of $\sqrt{b_i}$. It should be clear, however, that it is a nontrivial task to correctly identify the appropriate test to accompany code that is removed from a loop.

It is our feeling that the potential problems arising from optimizations of this type are of sufficient magnitude that no compiler can be constructed to properly transform code which is not strictly safe. The code shown in this example is poorly written. However, programmers do write code containing loop constant expressions, some of which are quite time-consuming.

Our feeling on the matter is that the compiler should report the existence of loop constant expressions to the programmer whenever a strictly safe transformation is not possible. The programmer would then have the option of recoding the loop if he felt the transformation to be worthwhile.

11.2.3. Global and Local Variables

Example 11.3

A problem of a different type, which should be discussed here, is concerned with so-called global and local variables. A global variable is a variable whose contents are accessable to the entire program, while a local variable is a variable which can be accessed only within the confines of a smaller program structure, say an ALGOL 60 block or a subroutine. In FORTRAN the global variables are the variables in COMMON or those variables which have been EQUIVALENCED to COMMON variables.

It is essential that attention be paid to the effects subroutines may have on expressions involving global variables. For example, assume that q is a global variable in the sequence

$$q \circ r \tag{11.3.1}$$

$$f(x, y, z) \tag{11.3.2}$$

$$q \circ r \tag{11.3.3}$$

We would like to establish whether the occurance of $q \circ r$ on line (11.3.3) is redundant or not. If q is defined in the subroutine f (or if any synonym of q is defined in f) or in any subroutine involved by f, then the expression $q \circ r$ is killed on line (11.3.2) and the expression is not redundant. Clearly, such an analysis is trivial, provided the code for f and the subroutines referenced by f are available to the compiler.

Unfortunately, some programming languages, including FORTRAN and JOVIAL, permit compilation of subroutines (called subprograms) at a different time than the main program. Such a language specification prohibits the complete data flow analysis required for redundancy analysis. Hence in these languages the assumption must be made that all expressions involving global variables are killed following any call on a subroutine that can be compiled independently, *even if the code for that subroutine is present at compile time.*

11.2.4. Other Language Restrictions

There are certain other problems in major programming languages which have similarly bad effects on data and control flow analysis, among which is the ON condition of PL/I. An optimizing compiler must take these restrictions into account, lest unaccountable errors occur in previously working programs.

We have mentioned the problem of certain optimization transformations actually slowing the flow of a program as opposed to increasing its efficiency. This phenomenon occurs at any time code is moved from a block b into a block b' which is executed more frequently than b. The condition is aggravated when code is moved from an arbitrary block into a point of confluence of the program, a move which assures the execution of a piece of code which might not otherwise be executed.

11.3. EXECUTION FREQUENCY ANALYSIS

The optimization strategy employed in Chapter 10 is based on the assumption that the code in a proper loop is more likely to be repeatedly executed than the code in the immediate back dominator of that proper loop, and so forth. Thus an attempt is made to move frequently computed expressions to a point in the program where their execution would be minimized.

Assume that by some means the relative execution frequencies were known for the basic blocks of a program. Then those expressions for which there existed a choice of basic blocks into which to move their computation could be placed into the appropriate lower-frequency blocks and some saving would result in execution time. There would also be an indication of loops which could be either unscrolled or fused, and the analysis of code in blocks with low execution frequency could be minimized without adverse effects on program performance. There would also exist reasonable criteria for choosing between subroutine linkages, register allocation schema, and storage allocation mappings.

Ideally, it is possible to obtain execution frequency information by tallying the number of times each basic block is entered over a number of program runs over different sets of data. This method is, of course, subject to the usual problems related to obtaining good statistical samples.

11.3.1. Execution Frequency Equations

A control flow graph possessing a unique entry point is analogous to other flow graphs in that a variant of Kirchhoff's law of conservation of flow is applicable. Let numbers f_i represent the execution frequency of the basic block i, and let p_{ij} be the probability of branching directly from node i to node j, so that $\sum_{j \in \Gamma i} p_{ij} = 1$. Then we have that

$$f_i = \sum_{j \in \Gamma^{-1} i} p_{ji} f_j + e(i) \qquad (11.4)$$

where $e(i)$ is the number of times node i is entered externally. Since the only node entered from a point external to the program is the unique entry node e, $e(i) = 0$ for $i \neq e$.

Hence if the system (11.4) is solvable, then given the transition probabilities (p_{ij}), the execution frequencies f_i can be determined up to a constant factor. Now (11.4) can be written in the matrix equation form

$$f = Pf + e$$

or

$$(I - P)f = e$$

where e is the vector $(n, 0, 0, \ldots, 0)$, $P = (p_{ij})$, n is the number of times the program is entered, and I is the identity matrix. Hence (11.4) is solvable if and only if $(I - P)$ is nonsingular and the methods of inverting sparce matrices are applicable.

Alternatively, given the transition probabilities (p_{ij}) it is possible to solve (11.4) by first partitioning the program graph into intervals and then solving the associated equations in a manner analogous to the solution of the redundancy equations of Chapter 8. The result of the first stage of that analysis will be a relative execution frequency for the blocks of each interval. If that information is sufficient for the optimization algorithms being employed, analysis may be terminated at that point. Otherwise, the derived graphs would be analyzed, etc.

The difficulty in the problem is not one of solving (11.4) but of obtaining reasonable values of the p_{ij}. It is apparant that even a minute difference in the transition probabilities for the back-latches of a loop can markedly affect the execution frequency of the loop. We are unaware of any reasonable solution to that problem.

We suggest that the interested reader refer himself to Knuth's study of FORTRAN programs for some interesting empirical results in this area.

12 SUBROUTINE LINKAGES

12.1. SUBROUTINES

Often the encoding of a program includes certain code sequences which occur in several distinct portions of the code and are either identical or parametrically identical to one another. Almost all programming languages provide one or more notational devices to reduce the tedium of actually writing such code in explicit form in those program segments where the code is needed. This notational aid is called severally by the names *subroutine*, *function*, *macro*, *procedure*, *subprogram*, etc. We shall normally refer to them simply as subroutines.

Subroutines provide numerous advantages to the programmer in addition to minimizing the amount of code he need write by hand. The functional notation involved permits a greater lucidity and conciseness in the expression of an algorithm and often enhances the programmer's insight into his problem. Some programming languages allow the writing of recursive subroutines, so that standard mathematical functions which are normally defined recursively can also be programmed that way. Examples include

$$
\begin{aligned}
0! &= 1 \\
n! &= (n-1)! \cdot n \quad (n \geq 1)
\end{aligned}
$$

and

$$
\begin{aligned}
f_{-1} &= 0 \\
f_0 &= 1 \\
f_n &= f_{n-1} + f_{n-2} \quad (n \geq 1)
\end{aligned}
$$

as standard definitions of the factorial function and the Fibonacci sequence.

12.2. SUBROUTINE LINKAGES

There is a cost associated with the use of subroutines, namely that of transferring the flow of control to the subroutine and then back again to the next instruction in the code sequence. There must also be a communication of the parameters used by the stereotyped code in the subroutine and any values to be returned to the main program by the subroutine. This process is generally referred to as *subroutine linkage*.

Many programmers consider the overhead costs of subroutine linkage to be so high that they limit their use of subroutines to a minimum. While traditional subroutine linkages normally require several instructions, it is not clear that the cost is as severe as is often feared, particularly as a percentage of the running time of the routine itself.

12.2.1. Closed Subroutine Linkages

Traditionally, a *closed subroutine linkage* is employed. The code on either side of the subroutine call is comprised of a *prologue* and an *epiloque*, while the subroutine is divided into a proloque, a subroutine body, and an epilogue.

As an illustration of these linkage structures, we shall treat a call on a subroutine f with three parameters: The first two are so-called input parameters, and the last an output parameter. [The input parameters are the values (or names of values) that the subroutine manipulates, while the output parameter is the name of a variable into which the subroutine is to store a result.] The call on the subroutine would appear in a sequence of the form

$$\begin{array}{c} \cdot \\ \cdot \\ \cdot \\ f(\alpha, \beta; z) \\ \cdot \\ \cdot \\ \cdot \end{array}$$

The prologue would include the evaluation of the expressions α and β and the storing of their values or information, i.e., placing their values into some conventional location. The prologue might also include information about the output parameter z. Finally, information is generated about the location of the instruction to which the subroutine is to return control, and a branch is made transferring control to the subroutine.

The epilogue, if present, would be composed of code to achieve storing the results of the subroutine in z, provided the subroutine does not perform that act directly.

The prologue and epilogue may also save and restore the contents of certain registers of the computer, thus preserving the program environment.

The subroutine prologue interprets the parametrical data and stores the location to which the subroutine is to return control according to convention, while the subroutine epilogue stores return data in a well-defined manner and returns control to the specified location.

12.2.1.1. Optimization of Closed Subroutines

It is quite difficult to optimize closed subroutines to any significant extent. It is possible that some of the expressions for the input parameters are either redundant or involve compile-time computations. If the subroutine does not return any values and if there is no necessity to restore the program environment through an epilogue following the return of control to the normal instruction sequence, it may occur that a minor branching optimization is possible. The situation we have in mind is the case where a subroutine call is followed immediately by an unconditional branch to some other program location. In such a case, the optimizing compiler would be able to provide the subroutine with the branch location instead of the branch statement as its return address, thus eliminating an unnecessary instruction.

We note in passing that we have generalized the closed linkage to the extent that many programming languages do not require or use both prologues and both epilogues and that, in many cases, some of the initialization we have assigned to the subroutine call prologue is actually performed in the subroutine prologue and vice versa. Nor have we gone into detail on the actual implementation of such language dependent variations as recursive calls and variables, subroutine entry points, or the like. While of definite interest to the compiler writer, we feel that they are too specialized for this treatment.

(For the remainder of this chapter we shall make the assumption that the code for all subroutines, functions, etc., is available to the optimizing compiler and that no subroutine may be compiled independently of the entire program. For those languages in which this assumption is invalid, the closed linkage is mandatory and the remainder of the discussion inapplicable.)

12.2.2. Open Subroutine Linkages

It is possible that the code required for a program written with a subroutine f is actually longer than it would be if the subroutine body were coded in full at each call to f. This is clearly the case if either f is called at most once or if the closed linkage is longer than the body of f.

If each call on a subroutine is replaced with the body of the subroutine into which the compiler directly substitutes all its parameters, we have what is called an *open encoding* of the subroutine. Of course, the compiler will continue to distinguish between global and local variables bearing the same "name," but otherwise the resulting code has all the appearance of a macro expansion.

12.2.2.1. Advantages of Open Encoding of Subroutines

There are numerous advantages to the open encoding over the closed linkage. Foremost among these is the absence of most of the prologue and epilogue code. (Some code to save and restore the program environment may remain necessary.) However, a greater potential saving is a result of the subroutine's code becoming one with the program, for now that the subroutine's code is visible in context to the optimizing compiler, it is possible to expose new compile-time computations (arising from constant arguments in the parameter list), to expose new redundant expressions, to eliminate inaccessable code (which arises in generalized subroutines), and, in those cases where the subroutine is contained in a proper loop, to apply reduction-in-strength techniques to expressions involving the induction variable.

12.2.2.2. The Use of Execution Frequency Information

If only the linkage is eliminated, the gain in efficiency may not be great, since the linkage may represent but a small percentage of the execution time for the subroutine or the program, but if the subroutine call lies in a portion of the program with a high execution frequency, even this gain may magnify to a worthwhile figure.

If it is possible to obtain reasonable execution frequency data, an interesting compromise of space and time optimization is possible. In those portions of the program with high execution frequency, the open encoding of subroutines could be exploited to advantage at some sacrifice to space, while in the areas which are rarely executed, little is lost using the traditional closed linkage. Hence a specific subroutine may appear in both open and closed linkages within a given program.

12.2.3. Semi-Open and Semi-Closed Subroutine Linkages

Certain subroutines are written in an extremely general manner to permit the processing of numerous approximately similar situations. For example, a compiler may contain a subroutine that loads an arbitrary variable into an arithmetic register. The routine's input parameters would identify the register into which the variable is to be loaded, identify the variable as a scalar; vector; matrix; multidimensional array; partial machine word; integer; floating-point, literal, or Boolean structure; etc. Then a series of decision points within the routine would produce the appropriate code (including necessary indexing) to locate the entity and move it into the designated register.

Often, a number of the input parameters to such a general-purpose subroutine are constants. Indeed, the routine that fetches data might very well be used in a different mode to store data. The distinction could be made if one of the parameters were, say, 0 or 1.

A reasonable data and control flow analysis of the program and the designated subroutine might indicate that two successive calls on the routine differ in at most a small number of input parameters and that the subroutine does not modify its input parameters. There are several potential optimizations open to the optimizing compiler in such a case, the simplest of which is to modify only those input parameters of the second subroutine call which differ from the first.

Analysis may well indicate that the subroutine is actually composed of two or three independent subroutines (in the sense that whenever any one part is executed, the remainder is skipped) and that the determination of which part is to be executed is based on the value of an input parameter which is always set to a scalar constant. In such a case it is interesting to consider the consequences of actually regarding the subroutine as a set of independently accessible subroutines.

It is often the case that not all the parameters for the original subroutine are used by any one of its component subroutines. Instead, each of the component subroutines uses a proper subset of the input parameters and ignores the rest. This is often indicated by a subroutine call in which many parameters are 0.

In the traditional closed linkage, each of the input parameters must actually be set, even though the executed portion of the subroutine references only a few of them. When this subroutine is broken down into its component subroutines, the compiler can determine the input parameters for each component subroutine and then, at each call point, produce code which sets only those parameters.

Each of the component subroutines can then be analyzed in the same manner and conceivably be broken down into smaller subroutines. The result is shorter closed linkages to the indicated subroutine and slightly faster execution due to the elimination of a number of unnecessary decisions that, in actuality, are compile-time computations.

The latter analysis produces a *semiopen* or *semiclosed* linkage. In the semiopen linkage, one further step is teken toward optimization: The subroutine body is scrutinized for the presence of code which is independent of the subroutine parameters. This code, if present, is moved out of the subroutine body and into the subroutine call prologue. Once there, it can be examined for common subexpressions and, in general, treated as part of the calling program.

12.3. SUMMARY

The open encoding of a subroutine produces the greatest opportunities for significant improvements in program efficiency and is markedly superior, in terms of execution time, to either the semiopen or closed subroutine linkages. However, in the latter cases, the enhancement will not be comparable to the code a competent programmer could have produced with diligent effort. It is doubtful that the average gain in efficiency from the semiopen linkage is worth the cost of the analysis, save in those program areas of the highest execution frequency.

EXERCISE

Produce an algorithm which yields the set of global variables and expressions which are killed by invoking a given subroutine (and the subroutines it calls, etc.). Modify the algorithm so that it takes account of the point of call for those subroutines with variable entry points.

13 THE ELIMINATION OF DEAD CODE AND THE ALLOCATION OF STORAGE AND REGISTERS

13.1. DEAD COMPILER VARIABLES

In Chapter 9 we presented algorithms for identifying those variables which are busy on exit from a block. We concluded that any variable not busy on exit from a block in which it is defined is a dead variable and that the store operation should not be performed. Similarly, if α is a compiler variable (arithmetic expression) which is not busy on exit from the block in which α is defined, then α is a dead variable (expression).

It is interesting to ponder how α could be a dead variable, since α either was present in the original program or was introduced to the program by the compiler. The simplest way for this to occur would be a coding situation wherein some programmer variable v is given the value of a fresh computation of α and that definition of v is found to be dead. If, after removing the reference to α in the deleted definition of v, it is found that there are no α-busy paths from that computation of α, then the value of α ought not to be stored in the variable α, and the computation of α ought to be deleted. Now if α is defined in terms of some prior computation β, the same analysis may eliminate β, and so forth.

13.1.1. The Distinction Between Compiler Variables and Arithmetic Expressions

We have been identifying α as both a compiler variable and an arithmetic expression. The code sequence shown below should clarify the distinction between interpretations:

$$\alpha_1: \quad q \circ r \tag{13.1}$$

$$\alpha_2: \quad \alpha_1 \circ s \tag{13.2}$$

$$v \longleftarrow \alpha_2 \tag{13.3}$$

$$\alpha_3\colon \quad \alpha_1 * t \tag{13.4}$$

$$s \longleftarrow \alpha_3 \tag{13.5}$$

$$\vdots$$

$$\alpha_{10}\colon \quad \alpha_1 * \alpha_3 \tag{13.6}$$

Note that α_2 is killed on line (13.5). Hence there is no reason to store the value of α_2 in the variable α_2. On the other hand, α_1 and α_3 are each referenced subsequently and are (presumably) not killed prior to their use on line (13.6). Consequently, they should be stored somewhere so that they need not be recomputed on line (13.6). The storage location of, say, α_1 may be either a register or a memory location, pending on the availability of arithmetic registers and the number of intervening computations lying between the initial computation of α_1 and its final use.

Now, while it is necessary that a computation of α not be busy on exit from the block in which it is defined in order to be dead, the condition is not sufficient to eliminate all unnecessary stores. Consider the following code:

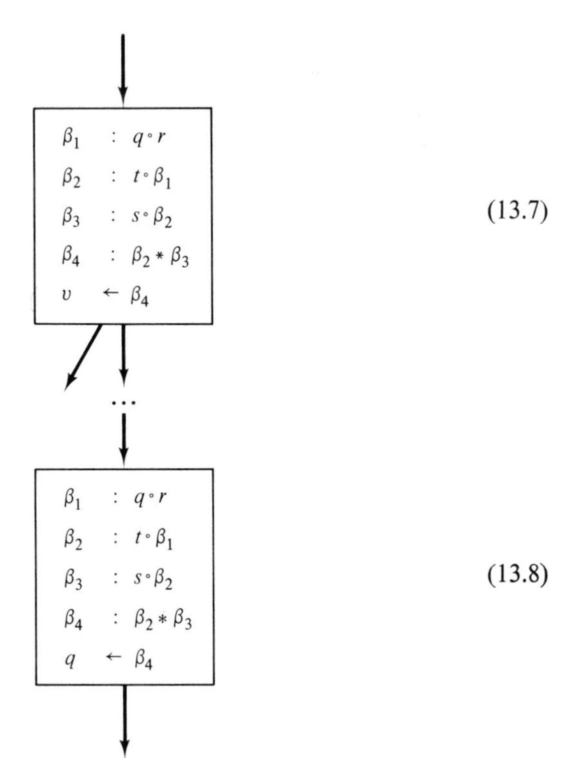

Here we assume that β_4 is available on entrance to (13.8) and, a fortiori, that β_1, β_2, and β_3 are available on entrance to (13.8). Clearly, there is no reason for a compiler to produce code for any of $\beta_1, \ldots, \beta_4$ in (13.8) provided at least β_4 has been saved in block (13.7). Note that each of $\beta_1, \ldots, \beta_4$ in (13.8) is redundant, provided at least β_4 has been saved in block (13.7). Note that each of $\beta_1, \ldots, \beta_4$ is busy on exit from (13.7). If we assume that the only subsequent use of the β_i is in (13.8), it is clearly unnecessary to store either β_1, β_2, or β_3 in (13.7), since, in a sense, they are dead.

13.1.2. Identifying Dead Compiler Variables

The answer lies in the construction of the code sequence that defines β_4. The programmer's code was of the form

$$v \longleftarrow (t \circ q \circ r) * (s \circ t \circ q \circ r) \tag{13.9}$$

which produces a tree resembling

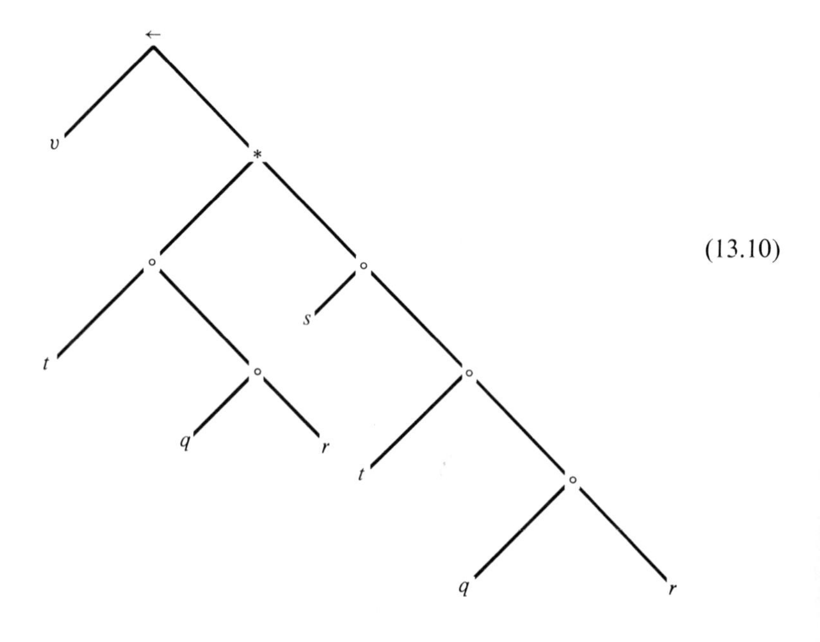

or

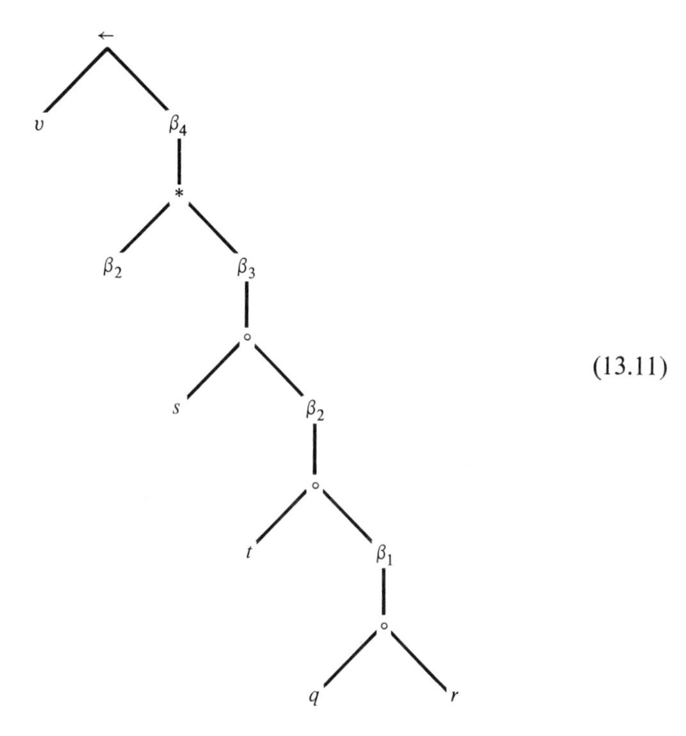

(13.11)

The tree (13.11) identifies the arithmetic expressions with their associated variables.

Now the tree for the code in block (13.8) is homeomorphic to (13.11) (except $v \longleftrightarrow q$), but we note that β_4 is available on entrance to (13.8), and consequently no further processing is required to define q. Hence it suffices to consider each assignment statement as a tree, with all subexpressions shown either explicitly or, if computed at a lower level of the tree, implicitly by name.

By construction, it follows that if the right-hand branch of the tree of an assignment statement is formally available on entrance to a block b, then it will actually be available on entrance to b provided it is stored into its associated compiler variable in the appropriate predecessor blocks of b.

13.1.3. Functional Expressions

Assume that we have an assignment statement in the block b of the form

$$w \longleftarrow \alpha_n$$

and that α_n is not available on entrance to b. Then let a scan be made down the tree considering each of the subexpressions which comprise α_n. The scan will terminate at the leaves of the tree or at those nodes representing expressions formally available on entrance to b (or already computed within b and not yet killed). The scan may be terminated at those nodes corresponding to available expressions, since for an expression to be available, so must be its constituent parts.

We shall therefore mark as *functional* those higher-level expressions in b which are formally available on entrance to b, and we say that it is *practicable* to store a busy definition of a compiler variable α in a block b' if there exists an α-busy path $\mu[b', b]$ leading to a block b in which α is functional. Note that the condition is stronger than busy on exit for one other reason: If $\mu[b', b]$ is an α-busy path, α need not be available on entrance to b (since there may exist $c \in \Gamma^{-1}b$ in which α is killed).

13.1.4. Identification of Practicable Stores

The equations of Algorithm 9.6 may be modified to expose the practicable stores in each block by introducing a variable

$$F_b(\alpha) = \begin{cases} 1 & \text{if } \alpha \text{ is functional in } b \\ 0 & \text{otherwise} \end{cases}$$

and solving for

$$\hat{E}_b(\alpha) = \lim_{k \to \infty} \hat{E}_b^{(k)}(\alpha)$$

where

$$\hat{E}_b^{(k+1)}(\alpha) = \hat{E}_b^{(k)}(\alpha) + D_b(\alpha) \cdot F_b(\alpha) \cdot \sum_{c \in \Gamma b} \hat{E}_c^{(k)}(\alpha)$$

[$\hat{E}_b^{(0)}(\alpha)$ and $D_b(\alpha)$ correspond to $E_b^{(0)}(\alpha)$ and $D_b(\alpha)$ in Algorithm 9.6.]

Then it is practicable to store a computation into a compiler variable α if $P_b(\alpha) = 1$, where

$$P_b(\alpha) = \sum_{c \in \Gamma b} \hat{E}_c(\alpha)$$

(The details of the proof are left to the reader.)

13.2. ELIMINATION OF DEAD CODE

The elimination of dead code must be left as the last optimization to be performed, since the other transformations move and introduce new code into the program. It should be noted that if a store into a programmer variable v at a point p is dead (i.e., v is not busy on exit from p), then the expression, α, being stored into v may be a dead computation. If α is not busy on exit from p, then α should either be marked as nonfunctional or be eliminated if

α was already nonfunctional. This action may, in turn, have an effect on other computations within the program. (Indeed, some early investigators in global optimization found that all of the code in their test cases was being eliminated by their experimental compilers! This occurred because ultimately all of their computations ceased being referenced.)

13.3. STORAGE ALLOCATION CONSIDERATIONS

Some interesting savings in storage allocation can result from an analysis of those variables (or data structures) which are busy at mutually disjoint times in the program. When two variables have this property, it is possible to store them into a common location in the computer store. Such analysis can prove to be quite profitable, as witness the work of academician A. P. Ershov, whose ATLAS compilers produced code for computers with extremely small memories.

Some computers have instructions for manipulating consecutive words of memory. A minor time saving can be realized on such machines if variables which are referenced together (such as two induction variables in a loop) are stored consecutively so that they can be accessed in one memory cycle.

13.4. REGISTER ALLOCATION

The efficient allocation of arithmetic registers is more a local optimization problem than a global one. However, the information provided by global analysis on those variables and constants which are in heavy use over an interval or within a strongly connected subgraph of the program can be useful in reducing the number of registers reserved by the compiler for interface purposes. For example, if a register is normally reserved for indexing an array and a proper loop makes no reference to that array, the register can be used for other purposes. This analysis may be simplified by considering the compiler-generated displacements as variables and analyzing the portions of the program where they are not busy.

The papers by Beatty and Kennedy each contain interesting approaches to register allocation and register assignment. It should be noted, however, that both techniques involve a considerable amount of additional analysis at the terminal stage of compilation. The expected gain over conventional local register allocation strategies may be quite small.

Appendix

I SELECTED ALGORITHMS EXPRESSED IN APL\360

Concurrent with the writing of this book, we had occasion to encode numerous graph-theoretic algorithms for computer simulation. We chose the APL\360 time-sharing system for their initial implementation due to the relative ease of programming. We readily admit that the concise notation is somewhat lacking in clarity, especially to those readers who are unfamiliar with APL. Consequently, we recommend that the reader familiarize himself with the IBM *APL\360 User's Manual* prior to studying our code.

Much of the data in which we are interested appears in tabular form, e.g., Fig. 8.4 is a listing of the Boolean computed-downward and kill vectors. The function we employed as a primitive output mechanism for a two-dimensional Boolean structure is shown below as *BOOLΔPRINT*, whose input parameter *X* is the structure to be printed.

```
     ∇ Z←BOOLΔPRINT X
[1]    D←'01'
[2]    Z←(((1⍴⍴X),3)⍴3 DFT⍳1⍴⍴X),'|',D[1+X]
     ∇
```

Here we employ an IBM supplied function, *DFT*, to convert the vector, *ι1ρρX*, of consecutive integers from 1 to the number of rows to be printed into a literal vector of right-justified integers with intervening blanks. This vector is restructured as a matrix which is laminated onto the literal equivalent of *X*, with a vertical bar placed in between.

Similarly, we employ the function *PRINT*, shown below, to show square matrices with column and row headers:

```
     ∇ Z←PRINT X;Q
[1]    '    ',3 DFT⍳¯1↑⍴X
[2]    ((Q,3)⍴3 DFT(⍳Q←1⍴⍴X)),'|',3 DFT X
[3]    '_________'
     ∇
```

APL\360 greatly facilitates initializing the connectivity matrix C̲ with which we represent the program graph. Below we show a listing of the program *SETC*, with which this is accomplished, together with a sample execution of *SETC* and *PRINT*. For convenience we show the graph as Fig. A.1.

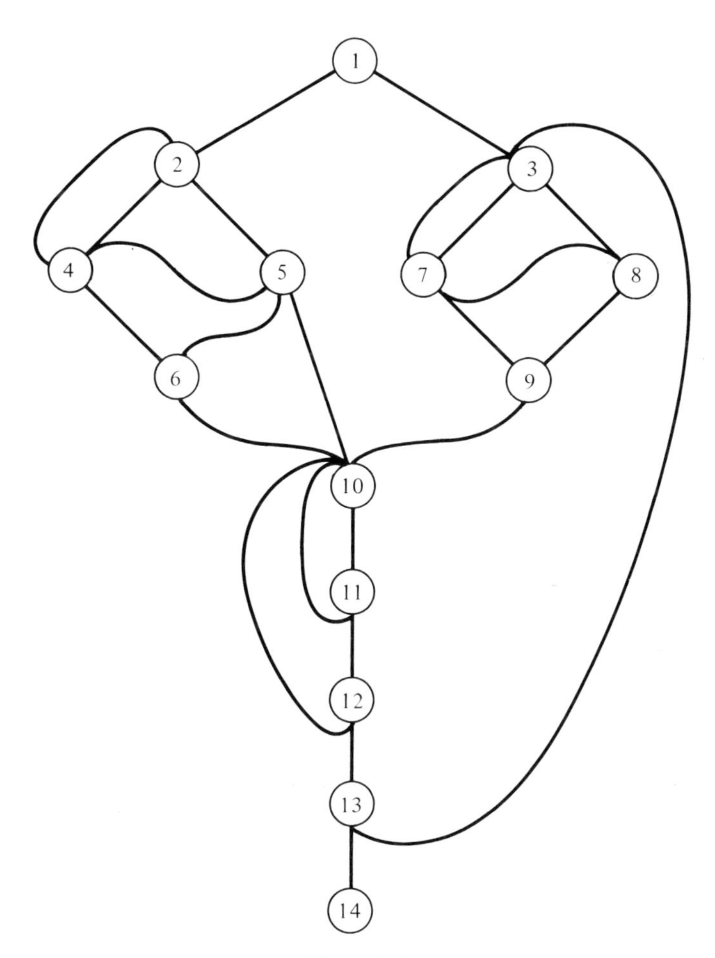

Figure A.1

```
      ∇ SETC;I;L
[1]     'INPUT NUMBER OF NODES'
[2]     C̲←(N,N←⎕)⍴0
[3]     I←1
[4]    LOOP:'SUCCESSORS OF NODE ';I
[5]     →(0=L←,⎕)/SKIP
[6]     C̲[I;L]←1
[7]    SKIP:→(N≥I←I+1)/LOOP
      ∇
```

```
      SETC
INPUT NUMBER OF NODES
⎕:
      14
SUCCESSORS OF NODE 1
⎕:
      2 3
SUCCESSORS OF NODE 2
⎕:
      4 5
SUCCESSORS OF NODE 3
⎕
      7 8
SUCCESSORS OF NODE 4
⎕:
      6 2
SUCCESSORS OF NODE 5
⎕:
      4 6 10
SUCCESSORS OF NODE 6
⎕:
      10
SUCCESSORS OF NODE 7
⎕:
      8 9 3
SUCCESSORS OF NODE 8
⎕:
      9
SUCCESSORS OF NODE 9
⎕:
      10
SUCCESSORS OF NODE 10
⎕:
      11
SUCCESSORS OF NODE 11
⎕:
      10 12
SUCCESSORS OF NODE 12
⎕:
      10 13
SUCCESSORS OF NODE 13
⎕:
      14 3
SUCCESSORS OF NODE 14
⎕:
      0
```

PRINT C

	1	2	3	4	5	6	7	8	9	10	11	12	13	14
1 |	0	1	1	0	0	0	0	0	0	0	0	0	0	0
2 |	0	0	0	1	1	0	0	0	0	0	0	0	0	0
3 |	0	0	0	0	0	0	1	1	0	0	0	0	0	0
4 |	0	1	0	0	0	1	0	0	0	0	0	0	0	0
5 |	0	0	0	1	0	1	0	0	0	1	0	0	0	0
6 |	0	0	0	0	0	0	0	0	0	1	0	0	0	0
7 |	0	0	1	0	0	0	0	1	1	0	0	0	0	0
8 |	0	0	0	0	0	0	0	0	1	0	0	0	0	0
9 |	0	0	0	0	0	0	0	0	0	1	0	0	0	0
10 |	0	0	0	0	0	0	0	0	0	0	1	0	0	0
11 |	0	0	0	0	0	0	0	0	0	1	0	1	0	0
12 |	0	0	0	0	0	0	0	0	0	1	0	0	1	0
13 |	0	0	1	0	0	0	0	0	0	0	0	0	0	1
14 |	0	0	0	0	0	0	0	0	0	0	0	0	0	0

Clearly, the graph of Fig. A.1 is not ordered according to the SNA. The graph must first be partitioned into intervals. Before listing the involved programs, we show their hierarchy in relation to solving the associated redundancy equations (Fig. A.2).

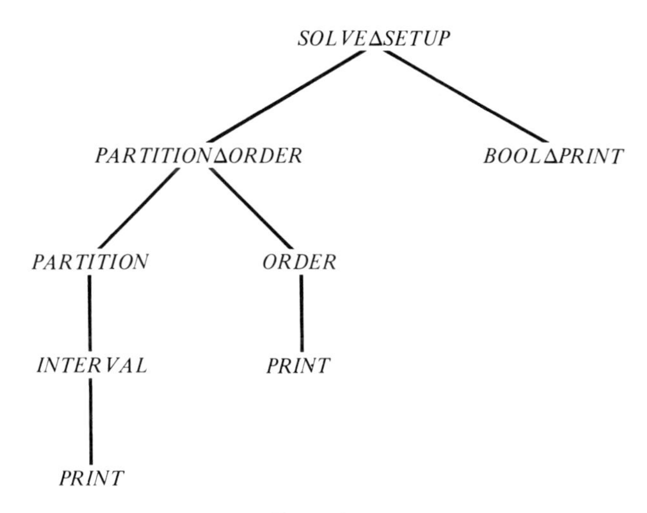

Figure A.2

```
     ∇ PARTITION∆ORDER;I;L
[1]    PARTITION C
[2]    L←ρI←1
[3]   LOOP:L←L,I ORDER F[I]
[4]    →(M≥I←I+1)/LOOP
[5]    C←C[L;L]
[6]    P←P[L;]
[7]    PRINT C
     ∇

     ∇ PARTITION C;E;I;Q;Z
[1]    E←Nρ1
[2]    F←,1
[3]    P←C INTERVAL(1,N)ρ1,(N-1)ρ0
[4]   LOOP:→(∧/0=E←E∧~Q←∨/P)/OUT
[5]    F←F,Z←(E∧∨⌿Q⌿C)/⍳N
[6]    →LOOP,0ρP←P,C INTERVAL Z∘.=⍳N
[7]   OUT:'THERE ARE ';M←¯1↑ρP;' INTERVALS:'
[8]    I←1
[9]   PRINT:'I(';F[I];')  :    ';F[I];'   ';((F[I]≠⍳N)∧,P[;I])/⍳N
[10]   →(M≥I←I+1)/PRINT
     ∇

     ∇ U←C INTERVAL Y;W;E;I;X;Z
[1]    E←0,(N-1)ρI←1
[2]    U←⍳0
[3]   TOP:Z←Y[I;]
[4]   GROW:W←E∧∨⌿(X←Z)⌿C
[5]    →(X≠Z←Z∨W\Z∧.≥W/C)/GROW
[6]    U←U,X
[7]    →((1ρρY)≥I←I+1)/TOP
[8]    U←⍉((1ρρY),N)ρU
     ∇
```

```
      ∇ORDER[⎕]∇
    ∇ Z←J ORDER H;L;S;I;T;M;K
[1]    →(0=ρL←(P[;J]∧C̲[;H])/ιN)/WEAK
[2]    S←((ρL),N)ρ0
[3]    K←(ρL)ρI←1
[4]   MAJOR:M←T←,L[I]
[5]   MINOR:T←(M←(~M∊H)/M),(~T∊M←(,∨/C̲[;M])/ιN)/T
[6]    →((0≠ρM),0=ρL)/MINOR,FINAL
[7]    S[I;ιK[I]←1+ρT]←H,T
[8]    →((ρL)≥I←I+1)/MAJOR
[9]    S←S[⍋K;]
[10]   K←K[⍋K]
[11]   H←ρI←1
[12]  STRONG:H←H,(~L∊H)/L←S[I;ιK[I]]
[13]   →((ρK)≥I←I+1)/STRONG
[14]  WEAK:T←L←ι0
[15]   →((1=I)∨(I←+/P[;J])=ρ,H)/FINAL
[16]   →MINOR,T←M←(,P[;J]∧,(((N,N)ρP[;J])∧C̲)∧.=(N,1)ρ0)/ιN
[17]  FINAL:Z←H,T
    ∇
```

In these programs we have the following global variables:

N = number of nodes in the graph

$\underline{C}$ = connectivity matrix of the graph, with $(N, N) \equiv \rho\underline{C}$

M = number of intervals in the graph

P = characteristic partition matrix, with $(N, M) \equiv \rho P$; i.e., $P[\,;I]/\iota N \equiv$ the indicies of the nodes in the Ith interval of the graph

F = vector of interval heads; i.e., $F[I]$ is the head of the Ith interval of the graph

Hence, *PARTITIONΔORDER* first calls *PARTITION*, which produces M, F, and P. It then produces an order vector L of the intervals, each ordered by *ORDER* on lines [3] and [4]. Finally $\underline{C}$ and P are permuted by L to correspond to the SNA order.

PARTITION employs a characteristic vector E of nodes not yet in any interval and a characteristic vector Q of nodes belonging to some interval. From these, a matrix $Z \circ .= \iota N$ is constructed such that each of its rows is the characteristic vector of the head of a new interval. [Inspection will reveal that when Z is constructed on line [5] of *PARTITION*, the vector Z is precisely $\Gamma^{\S}(I_1 \cup I_2 \cup \cdots \cup I_j)$, assuming J intervals have been constructed.] The process concludes when every node of the program graph belongs to some interval.

The function *INTERVAL* is a literal encoding of Algorithm 3.8. Starting with the first row of Y (which is $Z \circ . = \iota N$), the loop on lines [4] and [5] continues until X, the characteristic vector of the Ith interval on the previous iteration, is identical to the characteristic vector Z, which is the union of X and those nodes not in X all of whose immediate predecessors are already in X.

ORDER takes as arguments the index to the characteristic vector of the interval being ordered and the node heading that interval. *ORDER* produces a vector, Z, of the nodes in strict order. Line [1] determines the existence of a strongly connected component and, if none exists, branches to the label *WEAK*. Otherwise a matrix, S, is constructed from the vector, L, of back-latches such that the Ith row of S is an ordered chain leading from the interval head, H, to the back-latch $L[I]$. This ordering is achieved on line [5] by constructing vectors, T, of these chains. (The vector K lists the number of nodes in each chain.) A vector, H, is then constructed (lines [12] and [13]) from the rows of S according to the SNA.

If there is a weakly connected component of the interval, it is then ordered by the code on lines [5] and [6] after M is defined (on line [16]) as the set of nodes in the interval having no immediate successors in the interval. The reader will find the algorithm to be quite straightforward if he tries following the code with a sample graph.

Finally, we list *SOLVEΔSETUP*, the program which invokes the above functions and randomly initializes the computed downward and kill vectors, which it then prints on the user's console (lines [4] and [5]):

```
    ∇ SOLVEΔSETUP;Z
[1]   PARTITIONΔORDER
[2]   COMP←COMP,COMP←(N,32)ρ0=5|?(N×32)ρ100
[3]   KILL←KILL,KILL←(N,32)ρ0≠4|?(N×32)ρ100
[4]   16ρ' ';'COMP';34ρ' ';'KILL'
[5]   (BOOLΔPRINT COMP[;ι32]),(((1ρρZ),2)ρ'  '),Z←BOOLΔPRINT KILL[;ι32]
    ∇
```

Below we show the result of invoking *SOLVEΔSETUP* on the graph shown in Fig. A.3; we show the graph with the revised order numbers produced by the algorithm. This graph is easily seen to be irreducible. To simplify matters in the sequel, we shall delete the arc (13, 6) from the graph, so that it becomes fully reducible.

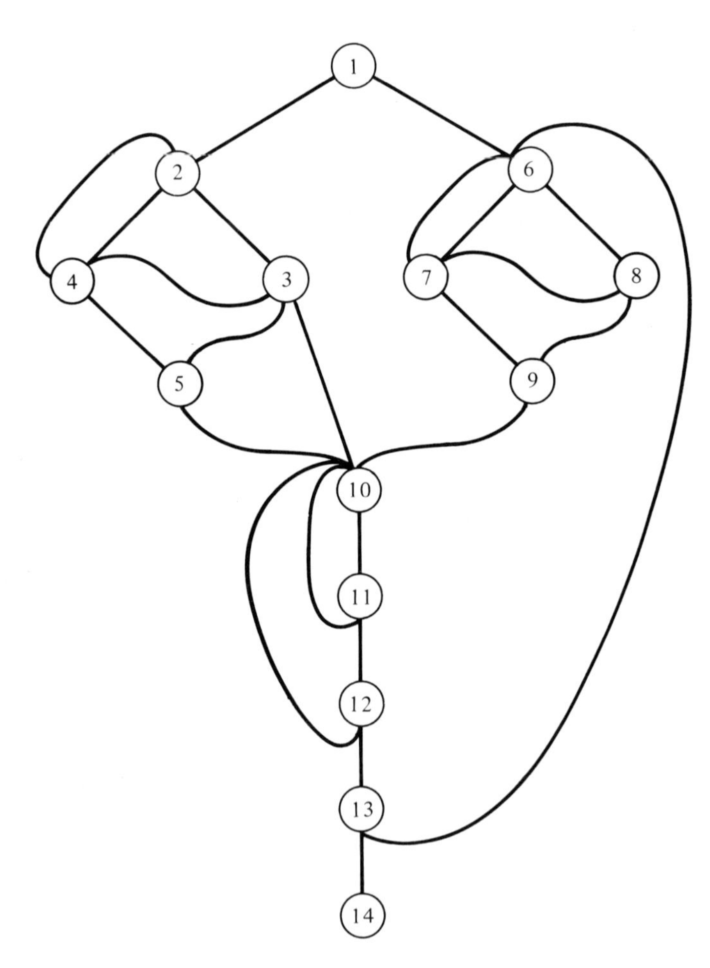

Figure A.3

```
          SOLVEΔSETUP
      THERE ARE 4 INTERVALS:
      I(1)   :     1
      I(2)   :     2   4   5   6
      I(3)   :     3   7   8   9
      I(10)   :     10   11   12   13   14
               1   2   3   4   5   6   7   8   9  10  11  12  13  14
           1|  0   1   0   0   0   1   0   0   0   0   0   0   0   0
           2|  0   0   1   1   0   0   0   0   0   0   0   0   0   0
           3|  0   0   0   1   1   0   0   0   0   1   0   0   0   0
           4|  0   1   0   0   1   0   0   0   0   0   0   0   0   0
           5|  0   0   0   0   0   0   0   0   0   1   0   0   0   0
           6|  0   0   0   0   0   0   1   1   0   0   0   0   0   0
           7|  0   0   0   0   0   1   0   1   1   0   0   0   0   0
           8|  0   0   0   0   0   0   0   0   1   0   0   0   0   0
           9|  0   0   0   0   0   0   0   0   0   1   0   0   0   0
          10|  0   0   0   0   0   0   0   0   0   0   1   0   0   0
          11|  0   0   0   0   0   0   0   0   0   1   0   1   0   0
          12|  0   0   0   0   0   0   0   0   0   1   0   0   1   0
          13|  0   0   0   0   0   1   0   0   0   0   0   0   0   1
          14|  0   0   0   0   0   0   0   0   0   0   0   0   0   0
      ---------
```

```
              COMP                                      KILL
 1|10101000000000110000000010010101       1|01011110001111010111010111101101
 2|00000001001011000000000100000101       2|11111011110111111111110010111110
 3|00000011000001000100001110101010       3|11101011111001001111101111000111
 4|00100000000010010000100000000000       4|10110111111111111011011101101110
 5|00010000000100100000001000000000       5|11110111111111110111101111110111
 6|01100010010000001000000000000100       6|00011100100011111011110110010011
 7|00001110010000000101010000000001       7|01101110101111110010010111111001
 8|00010000000110000000000000000000       8|11111111111011111110000111110011
 9|00100000000110010000000001100000       9|11111101100111111111110101101111
10|00000001000110000010000011011001      10|10111110110111001111111111111110
11|00000100100001010100000000001000      11|11111111110001110101111101011011
12|00011100000000000000000100001000      12|00111011111110111100111011111011
13|01100000000010010000011001001001      13|11111111110111111011111110101010
14|01000001000100000000110000000001      14|11011011111110111001111111010111
```

To solve the redundancy equations which follow from *KILL* and *COMP*, we employ a program *SOLVE*. The dependency tree of *SOLVE* is shown in Fig. A.4.

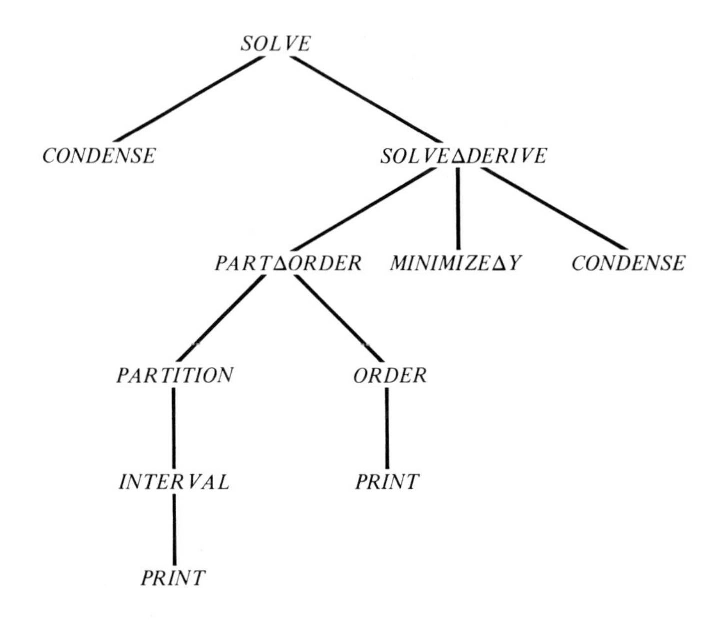

Figure A.4

```
     ∇ Z←P CONDENSE C;A
[1]    Z←((A←¯1↑ρP),(I←1)ρρC)ρ0
[2]  LOOP:Z[I;]←∨/P[;I]≠C
[3]    →(A≥I←I+1)/LOOP
[4]    Z←(Z∧⍉~P)∨.∧P
     ∇

     ∇ O←PARTΔORDER C̲;I
[1]    PARTITION C̲
[2]    O←ρI←1
[3]  LOOP:O←O,I ORDER F[I]
[4]    →(M≥I←I+1)/LOOP
     ∇

     ∇ X MINIMIZEΔY J;L;X1;X2;R
[1]    X1←((ρJ),32)ρ32↑X
[2]    X2←((ρJ),32)ρ¯32↑X
[3]    L←Y[J;]
[4]    Y[J;]←((L∧X1)∨R∧~X1),(X2∧L←L[;⍳32])∨(~X2)∧R←L[;32+⍳32]
     ∇

     ∇ SOLVE
[1]    D←'0≠1'
[2]    X←(64ρ0),[I←1]((M-1),64)ρ(32ρ1),32ρ0
[3]    Y←(N,64)ρ0
[4]  LOOP1:J←H←⌊/INT←P[;I]/⍳N
[5]    STRONG←⌈/INT
[6]    Y[H;]←COMP[H;]∨KILL[H;]∧X[I;]
[7]  LOOP2:→(STRONG<J←J+1)/EXIT2
[8]    →LOOP2,Y[J;]←COMP[J;]∨KILL[J;]∧∧/C̲[;J]≠Y
[9]  EXIT2:J←H
[10]   'INTERVAL ';I
```

```
[11] LOOP3:'EXIT ';(J<10)ρ' ';J;':   ';D[1+Y[J;ι32]+Y[J;32+ι32]]
[12]  →(STRONG≥J←J+1)/LOOP3
[13]  →((¯1↑ρP)≥I←I+1)/LOOP1
[14]  NODES←⍉P
[15]  G←ιM←ρF←F
[16]  N←1ρρY←Y
[17]  'THE EQUALITY VECTOR IS NOW:'
[18]  D[1+2×∧/Y[;ι32]=Y[;32+ι32]]
[19]  SOLVEΔDERIVE Q←(P←P) CONDENSE C
    ∇
```

```
    ∇ SOLVEΔDERIVE C;N;P;F;M;MAX;INT;I;J;K;Q;T;R;PRED;A;B
[1]  TOP:N←1ρρC
[2]   O←PARTΔORDER C
[3]   I←1
[4]   →(1=MAX←ρINT←,Q[⍋Q←OιP[;I]/ιN])/EXIT1
[5]   K←2
[6]  LOOP:A←∧⌿((PRED←C[;F[G[J]]])∧~P[;G[J←INT[K]]])⌿Y[;ι32]
[7]   →(0=+/B←PRED∧P[;G[J]])/MIN
[8]   A←A∧∧⌿B⌿Y[;ι32]
[9]  MIN:A MINIMIZEΔY NODES[J;]/ιN
[10]  →(MAX≥K←K+1)/LOOP
[11]  →EXIT1
[12] LOOP1:→(1=MAX←ρINT←,Q[⍋Q←OιP[;I]/ιN])/EXIT1
[13]  K←2
[14] LOOP2:(∧⌿C[;F[G[J]]]⌿Y) MINIMIZEΔY NODES[J←INT[K];]/ιN
[15]  →(MAX≥K←K+1)/LOOP2
[16] EXIT1:J←+/ρT←ρQ←(∨⌿NODES[P[;I]/ιN;])/ιN
[17]  'INTERVAL ';I
[18] LOOP3:'EXIT ';(K<10)ρ' ';K;' : ';D[1+Y[K;ι32]+Y[K;32+ι32]];(∨/Y[K;]≠Y[K←Q[J];])ρ'Δ'
[19]  →(T≥J←J+1)/LOOP3
[20]  →(M≥I←I+1)/LOOP1
[21]  →(1=ρG←G[F])/0
[22]  NODES←(⍉P)∨.∧NODES
[23]  Y←Y
[24]  C←P CONDENSE C
[25]  'THE EQUALITY VECTOR IS NOW:'
[26]  D[1+2×∧/Y[;ι32]=Y[;32+ι32]]
[27]  'CONTINUE? (Y/N)'
[28]  →('Y'=⍞)/TOP
    ∇
```

CONDENSE is a computationally efficient encoding of Theorem 7.13, which could be expressed more literally as

```
      ∇Z←P CONDENSE C
[1]    Z←(((⍉P)∨.∧C)∧⍉~P)∨.∧P
      ∇
```

However, the latter encoding is so inefficient on a large graph that it is more practical (and faster) to compute the derived graph by hand and input the new connectivity matrix at the keyboard. The iterative version of *CONDENSE* is practical for graphs with more than 100 nodes.

PARTΔORDER partitions and orders a graph, returning a vector of ordered indices for the straight order.

We need to explain *SOLVE* and *SOLVEΔDERIVE* in considerable detail. We shall start with *SOLVE*, which uses two matrices *X* and *Y* for available on entrance and available on exit vectors, respectively. Each of these vectors has 64 components. The first 32 correspond to the maximal solution of the equations, while the latter 32 correspond to the minimal solution.

Hence the matrix X is initialized with each row corresponding to an interval head, and we have $X[I;\iota 32] \leftarrow 32\rho 1$ for $I \geq 2$ and the remainder of X initialized to 0, respectively corresponding to the assumptions that the expressions are or are not available on entrance to the interval head.

On line [6], the availability on exit of each expression from the interval head, H, of the interval, I, is computed. Lines [7]–[8] then make the computation for the remaining nodes of the interval, in a sequence determined by the SNA. Note that

```
∧⌿C̲[;J]⌿Y
```

is an APL\360 idiom for the logical product

$$\prod_{d \in \Gamma^{-1} j} Y_d$$

since $\underline{C}[;J]$ is the characteristic vector of the immediate predecessors of node J, and $\underline{C}[;J]⌿Y$ is the matrix each of whose rows corresponds to the availability of exit vectors of one of the immediate predecessors of J.

The values of Y for the interval are output by lines [9]–[12], and if any unprocessed intervals remain, the branch on line [13] returns control to line [4].

Lines [14]–[19] initialize auxiliary variables for processing the derived graphs. These are

NODES	matrix whose rows are characteristic vectors of the nodes in the original graph for each derived node
$\underline{F}$	vector of interval heads in the original graph
$\underline{Y}$	value of Y from this iteration for comparison with Y in the next iteration
$\underline{M},\underline{N},\underline{P}$	remain M, N, and P for the original graph. M, N, P, and F will be used as defined previously for each derived graph

To describe *SOLVEΔDERIVE*, we need

DEFINITION **A.1**

Let $G = (X, \Gamma)$ be a finite directed graph and consider the derived sequence $(\mathcal{I}, \Gamma_{\mathcal{I}})$, $(\mathcal{I}^{(2)}, \Gamma_{\mathcal{I}^{(2)}})$, . . . , $(\mathcal{I}^{(k)}, \Gamma_{\mathcal{I}^{(k)}})$. An arbitrary derived node $i_n^j \in \mathcal{I}^{(j)}$ is the derived image of the nodes $\{x_{j1}, x_{j2}, \ldots, x_{jp}\} \subset X$. We call this set the *primitive ancestral nodes of* i_n^j. We call the unique node $x_{jt} \in \{x_{j1}, x_{j2} \ldots, x_{jp}\}$ such that either $\Gamma^{-1}x_{jt} \not\subset \{x_{j1}, \ldots, x_{jp}\}$ or $\Gamma^{-1}x_{jt} = \varnothing$

the *patriarchal ancestor of* i_n^j. The *primitive predecessors* of a derived node i_j are the immediate predecessors of the patriarchal ancestor of i_j.

SOLVE∆DERIVE is written as a loop and uses a matrix *C* to represent the derived graphs. The dimension *N* of *C* varies with the cardinality of the derived graphs. Hence lines [1]–[4] serve to initialize certain variables on each iteration of the program. In lines [4] and [12] the derived nodes of the *I*th interval of the derived graph are ordered according to the vector *O* which is set on line [2] and stored into a vector *INT*. If the first interval in the derived graph has but one node, then the primitive nodes comprising it are identical to the primitive nodes of the first interval in the previous derived graph, and as we have seen in Chapter 8, the redundancy equations for this derived interval have already been solved.

If the first interval contains more than one node, then the redundancy equations for all the primitive nodes comprising the first derived interval can be fully resolved, since the entry conditions for this interval are known. This is achieved in the loop on lines [6]–[10]. Here *A* is set to the logical product of the availability on exit vectors corresponding to the primitive predecessors of the patriarchal ancestor of the *J*th derived node, while *NODES* [*J*;] contains the primitive ancestors of *J*. *MINIMIZE∆Y* uses *A* as a mask with which to reconcile the minimal and maximal values of *Y*, as is described in the Example in Section 8.6.

The remainder of *SOLVE∆DERIVE* consists of a loop on lines [2]–[5] which further resolves the solution for the primitive ancestors of the nodes in the remaining intervals in the derived graph, less their individual interval heads, and a mechanism to initialize the appropriate variables so that the next derived graph can be processed, lines [21]–[28].

This simplified encoding of *SOLVE∆DERIVE* does not perform node splitting in irreducible graphs. However, the only modifications required would affect *CONDENSE*, which would perform the actual node splitting, and the detail around lines [21]–[22] which establish the primitive ancestral nodes in each interval.

We show below an execution of *SOLVE* (and *SOLVE∆DERIVE*) on our simple graph.

```
      SOLVE
INTERVAL 1
EXIT   1:  10101000000000110000000010010101
INTERVAL 2
EXIT   2:  ≠≠≠≠≠0≠1≠≠1≠11≠≠≠≠≠≠≠≠01≠0≠≠≠1≠1
EXIT   3:  ≠≠≠0≠011≠≠100100≠1≠≠≠01110101111
EXIT   4:  ≠01000≠1≠≠101101≠0≠≠100100≠0≠1≠0
EXIT   5:  ≠0≠100≠1≠≠11011000≠≠≠01100≠001≠0
INTERVAL 3
EXIT   6:  011≠≠≠10≠100≠≠≠≠10≠≠≠≠0≠≠00≠01≠≠
EXIT   7:  01101110≠100≠≠≠≠01≠1010≠≠00≠0001
```

```
EXIT  8:  0111≠≠10≠1011≠≠≠00≠0000≠≠00≠000≠
EXIT  9:  0110≠≠00≠0011≠≠100≠0000≠0110000≠
INTERVAL 4
EXIT 10:  ≠0≠≠≠≠≠1≠≠011≠00≠≠1≠≠≠≠≠11≠11≠≠1
EXIT 11:  ≠0≠≠≠1≠11≠000101010≠≠≠≠≠010110≠1
EXIT 12:  00≠111≠11≠000001010 0≠≠≠1010110≠1
EXIT 13:  011111≠11≠0010010000≠111010010≠1
EXIT 14:  010110≠11≠01100100001111010000≠1
THE EQUALITY VECTOR IS NOW:
10000000000000
THERE ARE 1 INTERVALS:
I(1)   :   1  2  3  4
INTERVAL 1
EXIT  1 : 1010100000000001100000000010010101
EXIT  2 : 101000001001011010000000010000001 01Δ
EXIT  3 : 10100011001001000100000111010111 1Δ
EXIT  4 : 10100000100101101000010010000010 0Δ
EXIT  5 : 10110001001101100000000110000010 0Δ
EXIT  6 : 01100010010011011000000010000010 1Δ
EXIT  7 : 01101110010011010101010100000000 1Δ
EXIT  8 : 01110010010111010000000010000000 1Δ
EXIT  9 : 01100000000111010000000010110000 1Δ
EXIT 10 : 00100000100011000000100001110110 01Δ
EXIT 11 : 00100101100001010100000101011001Δ
EXIT 12 : 00111101100000010100000101011001Δ
EXIT 13 : 01111101100010010000001110100100 1Δ
EXIT 14 : 01011001100110010000111101000000 1Δ
```

The transitive closure of a graph may be obtained from the program *TRANSITIVE*, shown below. Its output is shown for the graph of Fig. A.2 with (13, 6) deleted.

```
      ∇ TRANSITIVE;M;I;D;E;K
[1]     K←1
[2]     I←(⍳N)∘.=⍳N
[3]     E←D←I∨C
[4]   LOOP:E←E∨.∧D
[5]     →(N>K←K+1)/LOOP
[6]     'THE TRANSITIVE CLOSURE OF C IS:'
[7]     PRINT E
      ∇
```

```
         TRANSITIVE
   THE TRANSITIVE CLOSURE OF C IS:
           1  2  3  4  5  6  7  8  9 10 11 12 13 14
       1|  1  1  1  1  1  1  1  1  1  1  1  1  1  1
       2|  0  1  1  1  1  0  0  0  0  1  1  1  1  1
       3|  0  1  1  1  1  0  0  0  0  1  1  1  1  1
       4|  0  1  1  1  1  0  0  0  0  1  1  1  1  1
       5|  0  0  0  0  1  0  0  0  0  1  1  1  1  1
       6|  0  0  0  0  0  1  1  1  1  1  1  1  1  1
       7|  0  0  0  0  0  1  1  1  1  1  1  1  1  1
       8|  0  0  0  0  0  0  0  1  1  1  1  1  1  1
       9|  0  0  0  0  0  0  0  0  1  1  1  1  1  1
      10|  0  0  0  0  0  0  0  0  0  1  1  1  1  1
      11|  0  0  0  0  0  0  0  0  0  1  1  1  1  1
      12|  0  0  0  0  0  0  0  0  0  1  1  1  1  1
      13|  0  0  0  0  0  0  0  0  0  0  0  0  1  1
      14|  0  0  0  0  0  0  0  0  0  0  0  0  0  1
   ---------
```

We shall list below programs which establish the minimum distance between any two points in a graph and which explicitly list the nodes in any minimal path between two points in the graph. These routines, *DISTANCE* and *PATH*, use a distance matrix $\underline{D}$, in which $\underline{D}[I;J] = N^2$ if and only if there does not exist any path from I to J; otherwise $\underline{D}[I;J]$ is equal to the length of a minimal path.

```
      ∇ DISTANCE C;D;T;K;T;N
[1]     D←(N*2)ρ⌊(N←ρC)*2
[2]     T←,T←C
[3]     K←1
[4]    NEXT:D←((~T)\(~T)/D)+K⌊T\T/D
[5]     →(N≤K←K+1)/OUT
[6]     T←,T←T∨.∧C
[7]     →NEXT
[8]    OUT:⎕←D←(N,N)ρD
      ∇

      ∇ P←I PATH J
[1]     →((I=J)∨D[I;J]=1)/OUT
[2]     →((I>N∨D[I;J]≥N)/ERR
[3]     P←I,(((D[I;J]=D[I;]+D[;J])∧1=D[I;])⍳1) PATH J
[4]     →0
[5]    ERR:'THERE IS NO PATH FROM ';I;' TO ';J
[6]     →0
[7]    OUT:P←I,J
      ∇
```

Appendix

II THE PHASES OF AN OPTIMIZING COMPILER

II.0. INTRODUCTION

This Appendix describes the architecture of a specific optimizing compiler in which the order of the various analyses is based on pragmatic considerations of the algorithmic source language. A modification of the order of the analyses might be more appropriate for languages of a different type.

II.1. THE SYNTAX PHASE

Characteristically, it is in the scanning of a program's syntax that the source language statements are transformed into an intermediate language form which is logically equivalent to a tree. Higher-level language features are reduced to more primitive forms, generated labels and branches are produced, indexing expressions are expanded, subroutine linkages are indicated, etc. It is also in this phase that elementary diagnosis of programmer errors, such as using a variable prior to its initialization, may be produced.

There is a reasonable amount of preparation that must be performed in this phase for use by the optimizing compiler. In particular, the syntax pass may determine

1. The primitive binary expressions of the form $\alpha: v_1 * v_2$, where $*$ is some operator and v_1, v_2 are either variables or primitive expressions.

2. The most frequently occurring α expressions.

3. The basic blocks of the program graph.

II.2. THE GLOBAL ANALYSIS PHASE

The syntax phase differs only in minor respects from the syntax phase of a nonoptimizing compiler. Indeed, the few areas in which it differs tend to save some time in the optimization analysis phase and to prepare the intermediate language representation of the program as input for the code generation phase. It follows that, by construction, the second phase can be suppressed entirely should optimization not be desirable.

It is in the second phase that the bulk of analysis and optimization is performed.

II.2.1. Construction of the Graph

This portion of the second phase serves primarily to prepare the program graph for processing by the subsequent analysis passes over the data. Relatively local optimization may be performed at the same time.

II.2.1.1. Preparation of Data for Interblock Analysis

To start, the connectivity relations between the basic blocks of the program graph need be established.

Concurrently, bit-vectors for the most frequently occurring α expressions must be constructed to reflect those expressions which are killed and which are computed downward within each block. The associated vectors for busy on entrance (computed upwards) may also be initialized at this time.

II.2.1.2. Associated Intrablock Optimizations

Since the code in each basic block is being examined in this subphase, it is profitable to perform certain intrablock optimizations at the same time. These include

1. Folding (or *constant subsumption* or *constant propagation*).
2. Elimination of common subexpressions.
3. Early identification of inductive variables.

II.2.2. Partitions of the Graph

Following this initialization of basic information, the graph is partitioned into disjoint, maximal intervals. The nodes of each interval are ordered by the strict numbering algorithm (SNA). It is then possible to use this ordering to derive and employ the information described below.

II.2.2.1. Derivation of Pertinent Information

Starting with the first interval in the graph and proceeding in the strict order of the nodes, the solution is established for the redundancy equations; i.e., the availability on exit of each of the frequent α expressions is obtained for each basic block in the graph. This process requires passage to the derived graph sequence and may necessitate use of the node-splitting process.

Concurrent with the solution of the redundancy equation is the solution of the relative-frequency-of-execution equations.

Independent of the partition into intervals, the algorithms for busy on exit and very busy on exit can be run on either side of the solution of the other equations.

II.2.2.2. Utilization of Pertinent Information

The information obtained at this point is sufficient to eliminate from the program redundant subexpressions (i.e., those expressions which are available on entrance to a block and which are upwards exposed).

However, it is not necessary to actually perform the operation at this time, as it may be of interest to open short subroutines into the code sequence first. Indeed, it is here that an interesting conflict occurs which may well encourage our starting this subphase differently.

II.2.2.3. Diagnostic Analysis

Assume that the language does not permit the independent recompilation of subroutines and that some functions and procedures are capable of modifying data local to the basic block in which the subroutine is called. Then in order to maintain program equivalence, we assume that all α expressions involving any of the potentially modified variables were killed by the subroutine at the point of call. The global analysis just performed now indicates, for each of the subroutines, precisely which variables have actually been modified and hence which α expressions are actually killed.

This information may well depend on the existence of certain inaccessible nodes in subroutines which either explicitly kill data or which invoke other subroutines which in turn kill α expressions. These inaccessible blocks may be topologically inaccessible (indicating a programming error) or they may be inaccessible from a call in a block b due to the values of certain parameters, while they are quite accessible when the routine is invoked from some other block b'.

Having obtained this information, the kill and computed downwards vectors can be appropriately initialized to permit a more optimistic solution of the redundancy equations. The benefits of such analysis are most apparent in heavily subroutinized programs.

By analyzing the blocks in the order given by the SNA, it is possible to detect all cases where a variable *may* be used prior to its first definition. It is certainly possible to detect those cases where the variable is never (or always) defined prior to its first reference. However, there are some ambiguous cases, as is shown in Fig. A.5. If the path [e, a, b] is taken, v is undefined prior to its use, while if [e, a, b', c, a, b] is taken, there is no problem. (Such a graph may occur in a table-driven program.) The compiler should produce a warning for the programmer in such situations.

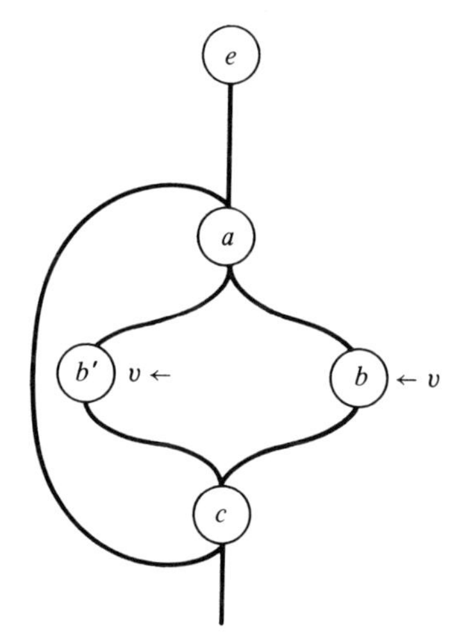

Figure A.5

II.2.3. Modifications of the Graph Based on Connectivity Considerations

The basic strategy involved in this global optimization process is to enhance the code contained in each suitably compact program loop, then the code of the suitably compact program loop containing it, and so forth outward until the entire program has been treated.

II.2.3.1. Identification of Proper Loops and Construction of Proper Loop Heads

To proceed in this manner, it is first necessary to identify the proper loops in each interval, to associate with each of them a unique proper loop head, and to modify the connectivity of the graph to reflect this change. Coincidentally, it is necessary to identify the points of confluence, back dominators, and articulation nodes of each interval and each maximal strongly connected subinterval.

II.2.3.2. Global Optimization

It is in this subphase that the majority of optimizations takes place. The typical transformations are briefly discussed below.

II.2.3.2.1. Loop Optimization. Here we start with the most deeply nested of the proper loops in the interval under consideration and move outward through the strongly connected subinterval and finally to the interval as a whole.

Code which is invariant within a proper loop is "moved" to the proper loop head provided that the code either occurs in a point of confluence of the proper loop or that the code is incapable of producing adverse side effects; e.g., such code should not have the potential of causing interrupts that it would not cause if left where it was originally, nor should its transposition cause the program to run less efficiently.

The inductive variables of the loop should be identified (the inductive variables in each basic block have already been found). The loop is then examined to identify possible situations which are ripe for reduction in strength.

Finally, loop range tests should be modified with an eye toward the elimination of otherwise dead loop variables.

II.2.3.2.2. Hoisting. The very-busy-on-exit information can be used to bring certain computations to a common block, provided that block is not located at an area of higher execution frequency. It is not only possible to conserve on space by hoisting, but also to improve execution efficiency in certain cases.

II.2.3.2.3. Subroutine Linkages. If the language specifications do not permit independent recompilation of subroutines, it is desirable at this point to choose between the closed, semiclosed, semiopen, and open subroutine linkages. Heuristically, it is advantageous to open short subroutines or at least to semiopen subroutines that are called in high-frequency loops.

Since the more open linkages usually expose a greater number of common subexpressions for possible elimination, etc., it may be necessary to reconsider folding and other elementary optimizations at the point of introduction.

II.2.3.2.4. Loop Modifications. Although it is possible to find short candidate loops for unscrolling at earlier phases of the analysis, it is preferable to delay such transformations until the loop analyses described above have reduced the size of all program loops. It is then possible to consider not only loops for unscrolling but also for fusion with other loops or even for identification with other loops. True, invariant expressions would be found as redundant subexpressions in an unscrolled loop, but it is doubtful that reduction in strength would still be applicable. The latter optimization is to be attempted whenever possible due to its power in reducing computation time.

As noted previously, when loops are unscrolled or fused, a rapid pass back through the code may produce new folding/common subexpression optimizations.

II.2.4. Reconsideration of the Graph

The majority of the optimizations having been performed, it is now time to prepare the graph for processing by the code generator. It is therefore necessary to indicate precisely which stores into α variables are to be performed, how storage is to be allocated, and which artificially introduced basic blocks are to become part of the program.

II.2.4.1. Branching Optimization

Consider the set of nodes possessing a unique immediate successor. (This set includes the artificially created proper loop heads.) If there exist nodes in this set which consist only of a branch to their successor, the appropriate branch in their predecessor nodes should be modified to branch to the successor node, and the node in question is deleted. Such nodes may correspond to generated labels in conditional statements.

Those nodes which branch to a unique successor which itself has a unique predecessor may be fused.

Since the code generator will process the nodes in strict order, it will always suppress one of the branches to an immediate successor node.

II.2.4.2. Dead Variable Analysis

Assignments to variables which are not subsequently used prior to redefinition are to be suppressed in this subphase. Included among these variables are the compiler-generated α variables. While the nonbusy programmer variables are easily detected for elimination, the case is different among the α variables since, under the regular definition of busy on exit, there may exist an α-busy path $\mu[b, b']$ in which α is not available on entrance to b' even though computed downward in b. That being the case, it may be desirable to eliminate the store into the α variable in b.

To accomplish this feat, the definition of busy on exit is modified to reflect availability on entrance, and busy on exit is recomputed for the program graph. Dead variable suppression then proceeds in a normal manner.

II.2.4.3. Storage Allocation

Those commensurate data structures which are busy at mutually exclusive times are given identical storage locations, thus reducing program size.

In algorithmic languages multidimensional arrays are given the core mapping that represents their most frequent usage, other than their declared representation.

Frequently used variables are given a register attribute to coerce their occupation of fast registers in areas of high usage.

II.3. THE CODE GENERATION PHASE

As we indicated, the code given this phase of the compiler may have had some optimization, no optimization, or total optimization performed on it. In all cases, the intermediate language form in which it is received by the code generator is identical. This phase will consequently employ standard local code optimization techniques on its input.

II.3.1. Register Allocation

Standard register allocation schema are employed, except that in certain portions of the program, certain variables or expressions may possess a register attribute. These are given preference for allocation to available registers.

II.3.2. The Window

Even the most sophisticated local code-generating techniques are capable of producing a store of a variable followed immediately by a load of the same variable. It has been found that the simple window device, through which a short series of consecutive instructions is viewed, has proved to be invaluable in terms of eliminating this frustrating inefficiency.

II.4. CONCLUSION

In this appendix we have briefly summarized the components of a specific class of global optimizing compiler. Indicated were optional modifications to its structure as well as the illustration of interdependencies between optimization analyses and transformations.

Appendix III EFFECTS OF PARTIAL RECOMPILATION ON THE GLOBAL OPTIMIZATION PROCESS

III.0. INTRODUCTION

In this appendix we shall systematically examine the effects of partial recompilation on a program that has previously been subjected to some form of global analysis and optimization and which is being recompiled due to some modification to the source code. In particular, we are interested in having to recompile and reanalyze as little of the program as possible while producing the best optimization possible under the constraints.

III.1. EFFECTS ON THE SYNTAX PHASE

Here we shall assume that the text of the original program is accessible by the compiler, that individual lines of code are identifiable by a relative sequence number, and that the submitted changes are identifiable under this scheme as additions, deletions, or modifications of existing statements.

The compiler should update the source language file to reflect these changes.

For each of the modifications local to some identifiable module of the program, sufficient context should be retrieved from the source file, parsed, and prepared for the global analysis phase. The relevant module and context will be derived below; for the moment, we prefer that it remain ambiguous.

III.2. EFFECTS ON THE GLOBAL ANALYSIS PHASE

As we assumed the existence of an accessible source file in the previous section, it becomes clear that in order to minimize analysis in this phase some information will be required from the original compilation of the program and that each successive partial recompilation will update that information. Exactly what that information is will be derived shortly.

III.2.1. Construction of the Graph

As in the complete optimizing compiler, we start by constructing a graph from the portions of the program under consideration and deriving the primitive data required by other subphases.

III.2.1.1. Preparation of Data for Interblock Analysis

The code is first broken into basic blocks. During this process folding and common subexpression suppression are performed while the computed downward and kill vectors are established.

The following possibilities exist:

1. No new blocks are introduced.
2. New blocks are introduced.
3. No kill vectors are changed.
4. Some expressions that were not previously killed now are.
5. Some expressions that were previously killed are not.
6. No computed downward vectors are changed.
7. Some expressions that were not computed downward now are.
8. Some expressions that were computed downward now are not.

There is no contradiction in assuming possibility 1 or 2; 3 or any of 4, 5; 6 or any of 7, 8. For example, 2 occurs with the addition of a conditional, loop, case statement or an unconditional transfer.

One can add a computation to a basic block without changing either the associated kill vector or the associated computed downward vector for the block:

Example:

Block before	Block after
$X =$	$X + Y$ $X =$

One can change both computed downward and kill for a block without explicitly modifying the code in the block:

Example:

$X + Y$
$f(a)$

where f did not originally modify X or Y but now modifies one or both of them.

Hence it is clear that the computed downward and kill vectors will have to be recomputed at least for the basic blocks in which code was modified and possibly in others as well, depending on our decision on the scope of our analysis on partial recompilations.

III.2.1.2. Associated Intrablock Analysis

Folding, common subexpression suppression, etc., may be carried out in the modified basic blocks as in the full compiler. Modifications to the kill vector by external code (functions, subroutines) must be given heed.

III.2.2. Partitions of the Graph

The addition of a basic block or a new branch is sufficient to modify the previous partition into intervals. Indeed, a new branch can change a reducible graph to an irreducible graph. Hence, so that any interblock analysis and/or optimization can be performed, the graph (or at least the part affected by the code modifications) must be repartitioned.

In light of this cost, it should be noted that if we wish only to perform intrablock analysis, the second phase of the incremental compiler will always perform only the optimizations of Section III.2.1.2 (with the assumption that external code can always be recompiled independently). This alternative must be considered, in that while the code produced under the assumption is suboptimal, the recompilation costs are minimal.

III.2.2.1. Derivation of Pertinent Information

Given that both the partition and the kill and computed downward vectors have most likely changed since the previous partial recompilation, we need to resolve the redundancy equations, busy on exit, very busy on exit, and execution frequency equations.

The effects of a code modification in a basic block has potential effects on each of the basic blocks accessible from it. (The influence is limited only by the presence of subsequent computations or annihilations of the concerned expression on all paths leaving the distinguished node.) In particular, a code modification in a basic block may affect the entry conditions into intervals distinct from the interval containing the modification.

Consequently, if this communication between intervals is capable of adversely affecting previously compiled code in unmodified intervals, then it is possible that all of the code in the program may need to be recompiled on each incremental compilation.

A simple alternative is to suppress interinterval communication. While this potentially reduces the interblock elimination of redundant subexpressions, only the interval or intervals containing modified code need be recompiled. Furthermore, because of the single-entry property of intervals, the newly compiled code can be patched into the existing program module with a minimum of work on the part of the code generator or link editor.

The implementation of this alternative simply entails the assumption that no data are available on entrance to any of the interval heads (or, equivalently, that all data are killed on exit from each interval). The result is that the minimal assumption on the redundancy equations is sufficient for the remaining analysis, and this is established without consideration of the derived graph sequence.

III.2.2.2. Utilization of Pertinent Information

Regardless of the assumptions that have been made about intercommunication between intervals, the interblock elimination of redundant code can be performed in a straightforward manner in this phase.

III.2.2.3. Diagnostic Analysis

Only analysis dependent on interinterval communication is affected by the decision on independent recompilation of intervals. Hence there is little effect on diagnostic analysis at this level. If all but exterior modules is recompiled, then the diagnostic analysis is practically unchanged.

III.2.3. Modifications of the Graph Based on Connectivity Considerations

Since the major differences between full global optimization and global optimization under partial recompilation constraints are most apparent when the interval is considered the basic unit of recompilation, we shall assume herein that the more restrictive assumption was made. In those cases where the modified module is being recompiled and there exist differences between the partial treatment and the complete treatment, they will be indicated explicitly, as applicable.

III.2.3.1. Identification of Proper Loops and Construction of Proper Loop Heads

The identification of proper loops and proper loop heads for the modified intervals is as before. However, the optimizations performed on the derived graph sequence must be curtailed due to the assumption that code in unmodified intervals is to remain unchanged.

The consequence of this restriction is that those loops and only those loops whose head and back-latches are contained in the same interval will be treated.

III.2.3.2. Global Optimization

In this section we shall treat the intrainterval optimizations specifically.

III.2.3.2.1. Loop Optimization. Within the affected interval, invariant code may be moved to the proper loop heads, code may be introduced into the proper loop heads to reduce computations within the loops which are not invariant, and reduction in strength of operators may be performed as usual.

III.2.3.2.2. Hoisting. As very busy on exit information only exists for the interval itself, hoisting will only affect expressions within the interval.

III.2.3.2.3. Subroutine Linkages. As subroutines are assumed to be independent external modules, their linkages may not be altered. However, common subexpressions in the standard calling sequence may be suppressed.

III.2.3.2.4. Loop Modifications. Within the interval under consideration, proper loops may be unscrolled.

Fusion of loops requires that the loops originally be disjoint and a fortiori, in distinct intervals. Hence fusion may not be performed if the interval is the basic program unit under consideration.

III.2.4.1. Branching Optimization

Since no code in an interval is potentially unreachable [all the nodes of $I(h)$ are by definition in $\hat{\Gamma}h$], the only possibilities here involve the identification of branches from nodes with a unique successor to nodes with a unique immediate predecessor in the same interval. Included in this family of nodes are those nodes which topologically possess several immediate successors but which have only one immediate successor due to data values at the decision point. The probability of detecting such nodes is only marginally decreased by the interval recompilation assumption.

III.2.4.2. Dead Variable Analysis

Dead variable analysis is not likely to bear as much fruit within one interval as when the entire submodule is considered, but since it was assumed that no communication exists between intervals, there will be no harm done to the program if the analysis is performed.

III.2.4.3. Storage Allocation

With the interval assumption, it is not possible to detect variables which can be overlayed.

With the submodule assumption, only local variables which cannot be affected by external modules may be considered, since it is only for them that the appropriate busy on exit information is available.

III.3. EFFECTS ON THE CODE GENERATION PHASE

Not surprisingly, the code generation phase is affected regardless of the partial recompilation assumption. Notably, to preclude regenerating unmodified code, some form of linkage must be provided to suppress execution of previously generated code and to transfer control to and from the modified code with minimal cost on execution efficiency.

This linkage is somewhat costly on the basic block level. The compiler keeps a current directory of the addresses of all basic blocks (expensive, since the number of basic blocks is large) and simply changes the first instruction in the original modified basic block to a branch to its replacement, duly modifying the basic block directory. The cost for the compiler is high, but little core space is lost by the modification.

For intervals, again a directory is maintained by the compiler. Only the first instruction in the original interval need be modified to transfer to its replacement, since intervals are single-entry. There is a lower compiler cost, but there may be a significant core loss.

For subroutines, linkage may be established by a vector mechanism wherein transfers are always directed to a table of branches and each subroutine occupies a unique slot. Then only the transfer address need be modified. The cost in terms of unused space can be minimized if all subroutine code is relocatable and a good link editor is utilized.

III.3.1. Register Allocation

The register allocation scheme must be modified to assume empty registers (except globally assigned base registers) at the start of each independently recompilable entity, be it a basic block, interval, or submodule.

III.3.2. The Window

The same restriction applies to the bands on the window. Note, however, that if the basic block is the unit of recompilation, at least one instruction must be left in each basic block to permit linkage in the event the basic block is subsequently modified.

III.4. CONCLUSION

We have indicated three distinct alternatives for the partial modification and recompilation of previously compiled and optimized programs. The idea has been to decrease compilation costs during a debugging phase while producing reasonably optimal code. Consequently, it is desirable that the compiler leave intact as much unmodified compiled code as possible on each recompilation.

The alternative requiring the least analysis recompiles only the basic blocks in which changes were made. It requires the least analysis but produces the least optimal code of the three alternatives as a consequence.

The second alternative treats each interval containing a basic block in which a modification was made. Analysis costs are still quite low, and a reasonable amount of the power of the optimizer can be exploited. In particular, some program loops can be fully optimized independently of the surrounding program environment.

The third alternative is the recompilation of the entire program submodule (subroutine, function, coroutine, access function, etc.) without consideration of external submodules. This alternative produces approximately the same optimization as the full global analysis described in Part II. The cost, however, is nearly as great (on a submodule basis) as full recompilation would be, the only difference being that other submodules are not considered.

We are personally of the opinion that the second alternative is the most cost/effective, since it is rare that a program in the debugging stage need be given long, costly runs. Rather, such programs are normally given sample data requiring minimal processing.

BIBLIOGRAPHY

GRAPH THEORY

BOOKS

C. BERGE, *Théorie des Graphes et ses Applications*, Dunod, Paris, 1958.

C. BERGE and A. GHOUILA-HOURI, *Programmes, Jeux et Réseaux de Transport*, Dunod, Paris, 1962.

R. G. BUSACKER and T. L. SAATY, *Finite Graphs and Networks: An Introduction with Applications*, McGraw-Hill, New York, 1965.

C. FLAMENT, *Applications of Graph Theory to Group Structure*, Prentice-Hall, Englewood Cliffs, N.J., 1963.

L. R. FORD, JR., and D. R. FULKERSON, *Flows in Networks*, Princeton University Press, Princeton, N.J., 1962.

F. HARARY, R. Z. NORMAN, and D. CARTWRIGHT, *Structural Models: An Introduction to the Theory of Directed Graphs*, Wiley, New York, 1965.

O. ORE, *Theory of Graphs*, American Mathematical Society Colloquim Publications, Vol. XXXVIII, Providence, 1962.

B. ROY, *Cheminement et Connexité dans les Graphes, Application aux Problèmes d'Ordonnancement*, Metra Spécial Série, No. 1, Paris, 1962.

PAPERS

D. G. CORNEIL and C. C. GOTLIEB, "An Efficient Algorithm for Graph Isomorphism," *J. ACM* **17** (1), Jan. 1970, pp. 51–64.

L. DIVIETI and A. GRASSELLI, "On the Determination of Minimal Feedback Arc and Vertex Sets," *IEEE Trans. Circuit Theory* **CT-15** (1), March 1968, pp. 86–89.

T. C. HU, "Revised Matrix Algorithms for Shortest Paths," *Siam J. Appl. Math.* **15** (1), Jan. 1967, pp. 207–218.

T. Kamae, "A Systematic Method of Finding All Directed Circuits and Enumerating All Directed Paths," *IEEE Trans. Circuit Theory* **CT-14** (2), June 1967, pp. 166–171.

A. Lempel and I. Cederbaum, "Minimum Feedback Arc and Vertex Sets of a Directed Graph," *IEEE Trans. Circuit Theory* **CT-13** (4), Dec. 1966, pp. 399–403.

T. C. Lowe, "An Algorithm for Rapid Calculation of Products of Boolean Matrices," *Software Age*, March 1968, pp. 36–37.

C. V. Ramamoorthy, "Analysis of Graphs by Connectivity Considerations," *J. ACM* **13** (2), April 1966, pp. 211–222.

J. C. Tiernan, "An Efficient Search Algorithm To Find the Elementary Circuits of a Graph," *Comm. ACM* **13** (12), Dec. 1970, pp. 722–726.

S. Warshall, "A Theorem on Boolean Matrices," *J ACM* **9** (1), Jan. 1962, pp. 11–12.

O. Wing and W. H. Kim, "The Path Matrix and Its Realizability," *IEEE Trans. Circuit Theory*, Sept. 1959, pp. 267–272.

GENERAL MATHEMATICS

G. Birkhoff, *Lattice Theory*, American Mathematical Society Colloquium Publications, Vol. XXV, Providence, 1960.

W. Feller, *An Introduction to Probability Theory and Its Applications*, Vol. 1, Wiley, New York, 1950.

J. G. Kemeny and J. L. Snell, *Finite Markov Chains*, Van Nostrand Reinhold, New York, 1960.

J. Riordan, *An Introduction to Combinatorial Analysis*, Wiley, New York, 1958.

H. J. Ryser, *Combinatorial Mathematics*, Carus Mathematical Monographs, No. 14. Mathematical Association of America, New York, 1963.

COMPUTER PROGRAMS AND OPTIMIZATION

BOOKS

A. V. Aho and J. D. Ullman, *The Theory of Parsing, Translation, and Compiling*, Vol. II: *Compiling*, Prentice-Hall, Englewood Cliffs, N.J., 1973.

J. Cocke and J. T. Schwartz, *Programming Languages and Their Compilers*, New York University, New York, 1969.

P. C. Goldberg, "Compilers," Chap. III in *Optimization Techniques* (unpublished), IBM, Yorktown Heights, N.Y., 1971.

K. E. Iverson, *A Programming Language*, Wiley, New York, 1962.

———, *APL\360-OS and APL\DOS User's Manual*, IBM Form Number SH20-0906-0, IBM, White Plains, N.Y., 1970.

R. RUSTIN [Ed.], *Design and Optimization of Compilers*, Courant Computer Science Symposium 5, Prentice-Hall, Englewood Cliffs, N.J., 1972.

J. T. SCHWARTZ, *Abstract Algorithms and a Set-Theoretic Language for Their Expression*, Courant Institute of Mathematical Sciences, New York, 1971.

PAPERS

A. V. AHO, R. SETHI, and J. D. ULMAN, "A Formal Approach to Code Optimization," *Proc. Symp. Compiler Optimization*, ACM SIGPLAN Notices **5** (7), July 1970, pp. 86–100.

F. E. ALLEN, "Program Optimization," *Ann. Rev. Auto. Programming*, **5**, 1969.

———, "Control Flow Analysis," *Proc. Symp. Compiler Optimization*, ACM SIGPLAN Notices, **5** (7), July 1970, pp. 1–19.

———, "A Basis for Program Optimization," *Proc. IFIPS Congress 71*, Ljublana, Aug. 23–28, 1971, pp. 64–68.

P. BACHMANN, "A Contribution to the Problem of the Optimization of Programs," *Proc. IFIPS Congress 71*, Ljublana, Aug. 23–28, 1971.

J. L. BAER and G. ESTRIN, "Frequency Numbers Associated with Directed Graph Representations of Computer Programs," Dept. of Engineering, UCLA.

J. C. BEATTY, "A Global Register Assignment Algorithm" IBM, San Jose, Calif., 1970.

V. A. BUSAM and D. D. ENGLAND, "Optimization of Expressions in FORTRAN," *Comm ACM*, **12** (12), Dec. 1969, pp. 666–674.

J. COCKE, "Global Common Subexpression Elimination," *Proc. Symp. Compiler Optimization*, ACM SIGPLAN Notices **5** (7), 1970, pp. 20–24.

J. COCKE and R. MILLER, "Some Analysis Techniques for Optimizing Computer Programs," *Proc. Second International Congress Systems Sci.*, Hawaii, Jan. 1969.

W. H. E. DAY, "Compiler Assignment of Data Items to Registers," *IBM Syst. J.*, **9** (4), 1970, pp. 281–317.

C. P. Earnest, "Some Topics in Code Optimization," Computer Sciences Corp., El Segundo, Calif., 1969.

C. P. Earnest, K. G. BALKE, and J. Anderson, "Analysis of Graphs by Strict Ordering of Nodes," *J. ACM* **19** (1), Jan. 1972, pp. 23–42.

A. P. ERSHOV, "ALPHA—An Automatic Programming System," Nauka, Novosibirsk, U.S.S.R., 1967.

———, "A Multilanguage Programming System Oriented to Language Description and Universal Optimization Algorithms," IFIPS Working Conference, ALGOL 68 Implementation, Munich, July 20–24, 1970.

———, "Parallel Programming," Computing Center, Novosibirsk 90, U.S.S.R., 1970.

———, "Theory of Program Schemata," *Proc. IPIP Congress 71*, Ljublana, Aug. 23–28, 1971.

W. GEAR, "High-Speed Compilation of Efficient Object Code," *Comm. ACM* **8** (8), Aug. 1965, pp. 483–488.

M. S. HECHT and J. D. ULLMAN, "Flow Graph Reducibility," TR 96, Computer Science Laboratory, Department of Elec. Eng., Princeton University, Princeton, N.J., 1971.

L. P. HORWITZ, R. M. KARP, R. E. MILLER, and S. WINOGRAD, "Index Register Allocation," *Research Paper RC-1264*, IBM, Yorktown Heights, N.Y., Aug. 1964.

IBM System/360 Operating System, FORTRAN IV (H) Compiler, *Program Logic Manual*, File No. S360-25(OS), Form Y28-6642-3, 4th ed., Nov. 1968.

R. M. KARP and R. E. MILLER, "Parallel Program Schemata," *J. Comp. Syst. Sci.* **3**, 1969.

K. KENNEDY, "Informal Paper on Register Allocation," Courant Institute, NYU, 1970.

D. E. KNUTH, "An Empirical Study of FORTRAN Programs," *Software-Practice and Experience*, **1** (2), 1971, pp. 105–134.

E. S. LOWRY and C. W. MEDLOCK, "Object Code Optimization," *Comm ACM* **12** (1), Jan. 1969, pp. 13–22.

D. C. LUCKHAM, D. M. R. PARK, and M. S. PATERSON, "On Formalized Computer Programs." *J. Comp. Syst. Sci.* **4**, 1970, pp. 220–249.

R. MILNER, "Equivalences on Program Schemes," *J. Comp. Syst. Sci.* **4**, 1970, pp. 205–219.

J. NIEVERGELT, "On the Automatic Simplification of Computer Programs," *Comm. ACM* **8** (6), June 1965, pp. 366–370.

R. T. PROSSER, "Applications of Boolean Matrices to the Analysis of Flow Diagrams," *Proc. Eastern Joint Computer Conference*, No. 16, 1959, pp. 133–138.

J. E. RODRIGUEZ, "A Graph Model for Parallel Computations," *Report Nos. ESL-R-398, MAC-TR-56*, Project MAC, Massachusetts Institute of Technology, Cambridge, Sept. 1969.

J. D. RUTLEDGE, "On Ianov's Program Schemata," *J. ACM* **11** (1), Jan. 1964, pp. 1–9.

J. T. SCHWARTZ, "Reduction in Strength (or Babbage's Difference Engine in Modern Dress)," IBM, Menlo Park, Calif., 1967.

———, "Various Graph-Theoretical Algorithms Related to the Optimization of Compiler-Generated Code" (draft), Courant Institute of Mathematical Sciences, New York, 1971.

G. S. SHEDLER and M. M. LEHMAN, "Evaluation of Redundancy in a Parallel Algorithm," *IBM Syst. J.* **6** (3), 1967, pp. 142–149.

G. S. TJADEN and M. J. FLYNN, "Detection and Parallel Execution of Independent Instructions," *IEEE Trans. Computers*, **C-19** (10), Oct. 1970. pp. 889–895.

J. D. ULLMAN, "Fast Algorithms for the Elimination of Common Subexpressions," TR 106, Computer Science Laboratory, Department of Elec. Eng., Princeton University, Princeton, N.J., 1972.

INDEX

V

W

Z